Mössbauer Spectroscopy

An Introduction for Inorganic Chemists and Geochemists

Mössbauer Spectroscopy
An Introduction for Inorganic Chemists and Geochemists

G. M. Bancroft
*Associate Professor, Department of Chemistry,
University of Western Ontario*

A HALSTED PRESS BOOK

John Wiley and Sons
New York

Published in the U.S.A.
by **Halsted Press**
A Division of **John Wiley and Sons, Inc., New York**

Library of Congress Cataloging in Publication Data
Bancroft, G. M.
Mössbauer spectroscopy.
"A Halsted Press book."
Bibliography: p.
1. Mössbauer spectroscopy. I. Title.
QC490.B35   537.5'352   73-3326
ISBN 0-470-04665-1

Copyright © 1973 McGraw-Hill Book Company (UK) Limited. All rights reserved. No part of this publication may be reproduced, stored in a retrieval system, or transmitted, in any form or by any means, electronic, mechanical, photocopying, recording, or otherwise, without the prior permission of McGraw-Hill Book Company (UK) Limited.

PRINTED AND BOUND IN GREAT BRITAIN

Contents

Chapter 1: Radioactivity, nuclear properties and nuclear gamma resonance — 1

   1.1 An introduction to radioactivity — 2
   1.2 Radioactive decay and the Heisenberg uncertainty principle — 4
   1.3 Nuclear properties — 5
   1.4 The Döppler effect — 7
   1.5 Resonant absorption and nuclear gamma resonance (the Mössbauer effect) — 8

Chapter 2: Mössbauer parameters and theory — 17

   2.1 The isomer shift — 17
   2.2 Quadrupole splitting — 21
         The EFG tensor — 23
         The sign of q — 28
         Separation of $q_{valence}$ — 29
         Units — 32
   2.3 Measurement of the sign of the quadrupole splitting — 33
   2.4 Quadrupole hyperfine parameters for $I > \frac{3}{2}$ — 35
   2.5 Magnetic splitting — 38
   2.6 Line shape, width, and area — 40
   2.7 Characteristics of a useful Mössbauer isotope — 41

Chapter 3: Experimental techniques — 47

   3.1 Basic Mössbauer equipment — 47
         The drive mechanism — 47
         Sources — 47
         Detectors and electronics — 49
   3.2 Other experimental methods — 50
         Absorbers — 52
         The cosine effect and source to detector distance — 53
         Calibration — 53
         A good spectrum — 57
         Background — 57
   3.3 Computational methods — 58
         The method — 59
         Criteria for a 'good fit' — 60
         Examples of the fitting procedure — 62

Chapter 4: Mössbauer spectroscopy as a fingerprint technique in inorganic chemistry — 68

- 4.1 Purity and characterization — 69
- 4.2 Detection of structurally different atoms in polynuclear compounds — 72
- 4.3 Solid-state decompositions — 77
- 4.4 The effect of temperature and pressure on the electronic structure of iron compounds — 80
- 4.5 Frozen solution studies — 83
- 4.6 Emission spectra — 86

Chapter 5: Centre shifts and bonding properties — 90

- 5.1 Correlation of centre shifts with electronegativities and σ and π bonding properties of ligands — 91
  - Iodine and xenon compounds — 91
  - Fe^{II}, Ru^{II} and Au^{I} compounds — 94
  - Sn^{IV}, Sb^{III}, and Fe^{II} high spin compounds — 97
  - 'Anomalous' Sn^{IV} centre shift results — 101
- 5.2 Partial centre shifts (p.c.s.) — 102
- 5.3 Correlation of centre shift values with the spectrochemical and nephelauxetic series — 107

Chapter 6: Bonding and structure from quadrupole splittings — 110

- 6.1 Correlation of quadrupole splitting with bonding properties of ligands and centre shift values — 111
  - Iodine compounds — 111
  - Xenon compounds — 114
  - Fe^{II} and Ru^{II} compounds — 116
- 6.2 Partial quadrupole splittings: the 2: −1 *trans*: *cis* ratio — 121
- 6.3 Derivation of p.q.s. values: rationalization of signs and magnitudes of quadrupole splitting — 127
  - Fe^{II} low spin compounds — 127
  - Co^{III}, Ir^{III} and Ru^{II} compounds — 130
  - Sn^{IV} compounds — 133
- 6.4 Structural and bonding predictions from p.q.s. — 139
  - Structure — 139
  - Bonding — 143
- 6.5 The effect of $q_{C.F.}$ — 145
  - Fe^{II} high spin — 146
  - Fe^{o} and π-Cp Fe compounds — 150
  - Au^{III} compounds — 151

Chapter 7: Mössbauer spectroscopy as a fingerprint technique in mineralogy and geochemistry — 155

- 7.1 Oxidation state and electronic configuration of iron in minerals — 156
- 7.2 Coordination number of iron in minerals — 162

Contents

7.3	Determination of the number of distinct structural positions and the assignment of peaks in complex spectra	165
	Pyroxenes	165
	Amphiboles	177
	Micas	184
	Epidotes	187
7.4	Line widths and areas	189
7.5	Correlation of quadrupole splitting with structural variations	190
7.6	Uses of characteristic Fe^{2+} and Fe^{3+} peaks in solid state processes	193
	Charge transfer spectra, colour and pleochroism	194
	Oxidation and weathering	196

Chapter 8: Quantitative site populations in silicate minerals   201

8.1	The Mössbauer method for determining site population ratios	202
8.2	Site populations and comparisons with X-ray and chemical analyses values	204
8.3	A survey of the method in amphiboles and pyroxenes	206
	The cummingtonite-grunerite series	206
	The orthopyroxene series	212
8.4	A possible geothermometer	215
8.5	Site populations and crystal field phenomena	221

Chapter 9: Mössbauer spectra of multi-phase assemblages   226

9.1	The method and general assumptions	228
9.2	Meteorites	229
9.3	Lunar soils and rocks	234
9.4	Terrestrial samples and a potential use in geochemical prospecting	238

To David and Catherine

Preface

In the last ten years, Mössbauer spectroscopy has made a great impact on a wide variety of scientific disciplines: many branches of chemistry and physics, biochemistry, the earth sciences and metallurgy to name a few. A number of books on Mössbauer spectroscopy have now been published (see the Bibliography at the end of the chapters) but no book has yet appeared which is intended for the *inorganic chemist* or *geochemist*. Yet these are two of the areas in which Mössbauer spectroscopy has made a very large contribution.

This book is an attempt to fill this void. It is intended as a basic introduction, and is specifically aimed at the senior undergraduate and postgraduate levels. Much of the material given in the book has been given in lectures at the Universities of Cambridge and Western Ontario at both levels. It is hoped that the book will also be useful both to scientists in related disciplines who are interested in a new technique, and to existing Mössbauer spectroscopists for research and teaching. Although a fairly large number of references are given (often in tables and figure captions), this book is *not* intended as a reference work.

The book is divided into three parts of similar length: Part 1 (chapters 1, 2, 3), the Introduction; Part 2 (chapters 4, 5, 6), Applications in Inorganic Chemistry; and Part 3 (chapters 7, 8, 9), Applications in Mineralogy and Geochemistry. The last two parts are self-contained and should enable the geochemist to omit Part 2, and the chemist to omit Part 3. In fact, the geochemist or mining engineer who is not particularly interested in, or adept at, theory could quite likely read just chapter 1, before skipping to chapters 7, 8 and 9. However, one of the main intentions of the book is to promote interdisciplinary contact between inorganic chemists and geochemists, and so it is hoped that both chemists and geochemists will peruse the whole book.

Because of the dual role of this book, the level of the book has been a great problem. The first three chapters present ideas basic to the theory and practice of Mössbauer spectroscopy. Wherever possible, simple calculations have been worked in detail to illustrate principles and magnitudes. Much of the background normally taken for granted in other publications is explained in these first chapters. In the quest for simplicity, I have deliberately avoided quantum mechanical treatments of such areas as the quadrupole Hamiltonian in chapter 2.

In chapters 4 through 9, a potential or recognized use of Mössbauer spectroscopy is stated, followed usually by a critical discussion of the advantages and disadvantages of Mössbauer spectroscopy in such an area. For example, in chapter 8, some considerable time is devoted to the assumptions and difficulties of the Mössbauer area ratio method before discussing uses of such area ratios in geochemistry.

At the end of each chapter (except chapter 9) a number of problems are given to further understanding, and to create interest in the growing body of Mössbauer papers. Answers to the problems, or references to the relevant papers are given at the end of the book.

Many colleagues and students have been extremely important in the conception, approach and production of this book. I am particularly indebted to Dr A. G. Maddock at the University of Cambridge. I began my PhD in September 1964 on the Mössbauer effect under his direction, and throughout my six years in Cambridge, his suggestions, criticism, and enthusiasm were of the utmost importance in the work which has led to the publication of this book. I am also particularly grateful to Dr R. G. Burns for introducing me to the field of mineralogy, and for his great effort and enthusiasm throughout our collaborative work. Those were exciting years.

Many other colleagues and students have contributed directly or indirectly to the publication of this book: Dr M. G. Clark, Dr M. J. Mays, Dr R. H. Platt and Dr A. J. Stone; and Mr K. D. Butler, Dr K. G. Dharmawardena, Dr R. E. B. Garrod, Mr E. T. Libbey, Dr B. E. Prater and Dr P. G. L. Williams. I would also like to recall the memory of a good friend and colleague, Dr L. O. Medeiros, who so tragically died before making what I am sure would have been an outstanding contribution to Mössbauer spectroscopy. I am also indebted to a great number of very competent typists both in Cambridge and London, Ontario.

Finally, I would like to express my greatest thanks to my wife Joan for her patience and understanding during the writing of this book.

Bancroft

1. Radioactivity, nuclear properties, and nuclear gamma resonance

The Mössbauer effect (or nuclear gamma resonance) was discovered in 1957 during R. L. Mössbauer's graduate work. The Mössbauer effect began to be widely applied to chemical problems after it was shown in 1960 that ^{57}Fe exhibited this resonant phenomenon, but the importance of the technique for mineralogical research was not realized until 1965. Since 1965, both the chemical and mineralogical interest in the technique has grown to the stage where it is becoming a routine research tool in many inorganic chemistry and mineralogy departments.

The great fraction of Mössbauer research has been carried out using ^{57}Fe and ^{119}Sn. However, over fifty other isotopes exhibit this effect, and at least twenty of these can give useful information without extreme experimental difficulties.

Mössbauer spectroscopy can be compared with infrared spectroscopy since both are basic resonant techniques employing a photon source, an absorber, and a photon detector. In infrared spectroscopy, the photon energy emitted by a hot filament is modulated by a grating, and photons of discrete energies can be resonantly absorbed by vibrating atoms. In Mössbauer spectroscopy, the gamma ray energy from a radioactive nucleus is modulated by imparting a Döppler velocity to the source, and the gamma rays of discrete energies can be resonantly absorbed by absorber nuclei. In both techniques, the number of transmitted photons are plotted versus photon energy, and a peak or peaks are observed where resonance occurs.

Before examining the Mössbauer theory in chapter 2, it is worthwhile to discuss concepts in the above paragraph which recur time and again in later chapters but may be unfamiliar to some mineralogists and inorganic chemists: radioactivity, nuclear properties, the Döppler effect, and nuclear resonant absorption.

1.1 An introduction to radioactivity

Essential to all Mössbauer experiments is a radioactive isotope which acts as a source of radiation in the Mössbauer experiment. It is desirable to give some background to the properties of the nucleus and the shorthand for representing nuclear reactions before discussing radioactivity in more detail.

Each nuclear species, or nuclide, is an assemblage of protons and neutrons. The atomic number Z is given by the number of protons. The number of neutrons is denoted by N, and the sum $N + Z$ gives the mass number A'. The symbol used to denote a nuclear species is the chemical symbol of the element with Z as a right subscript and A' as a left superscript. For example, $^{57}Co_{27}$ indicates that there are twenty-seven protons and thirty neutrons in this nuclide. Nuclides having the same atomic number but different mass numbers are called isotopes: e.g. $^{35}Cl_{17}$ and $^{37}Cl_{17}$ (usually abbreviated ^{35}Cl and ^{37}Cl). Atomic species having the same mass number but different atomic number are called isobars: e.g. $^{130}Xe_{54}$ and $^{130}Ba_{56}$. On the average, the elements with atomic numbers between 1 and 83 have more than three stable (non-radioactive) isotopes each. In addition to the stable isotopes, many elements (especially the heavier ones) have a number of isotopes which are radioactive, and all naturally occurring nuclides with atomic number greater than 83 (i.e., bismuth) are radioactive. These isotopes decay to other stable, or in turn, other radioactive isotopes by emitting one or more particles and/or photons. The most common types of decay are termed beta (β), alpha (α), and gamma (γ) decay.

Any radioactive decay process in which the atomic number Z changes, while the mass number A' remains unchanged is referred to as β decay. The three most common β decay processes are termed negatron (β^-) decay, positron (β^+) decay, and electron capture. In negatron decay, negative electrons are emitted, and the atomic number is increased by one unit. For example:

$$^{182}Ta_{73} \rightarrow {}^{182}W_{74} + {}^{0}\beta_{-1}. \tag{1.1}$$

On present views, this comes about by the transition of a neutron to a proton. The ^{182}W is formed in a nuclear excited state, and a number of gamma rays are emitted before it decays to the ground state.

The normal notation for nuclear reactions used in eq. (1.1) is analogous to those of chemical reactions. Both the mass number and the atomic number are conserved. In addition, properties such as charge, energy, and momentum are conserved. An electron is designated $^0\beta_{-1}$, a positron $^0\beta_1$, and a neutron $^1n_0$.

Positron, or β^+ decay, arises by the transition of a proton to a neutron. As seen in the following example, the atomic number decreases by one unit:

$$^{61}Cu_{29} \rightarrow {}^{61}Ni_{28} + {}^{0}\beta_1. \tag{1.2}$$

Radioactivity, nuclear properties, and nuclear gamma resonance

Again ^{61}Ni is formed in a nuclear excited state and a gamma ray is emitted on decay to the stable ground state.

Another way of decreasing Z by one unit as above is by electron capture. A neutron deficient or proton rich nuclide 'captures' an inner valence electron; the net result is a transition of a proton to a neutron. The most common Mössbauer isotope, ^{57}Fe, is formed by electron capture from ^{57}Co. The reaction may be represented as:

$$^{57}\text{Co}_{27} + {}^{0}\beta_{-1} \rightarrow {}^{57}\text{Fe}_{26}. \qquad (1.3)$$

The ^{57}Co, like a great many radioactive isotopes is produced by a nuclear reaction. In this case ^{56}Fe is bombarded with deuterons (^{2}H$_1$), and ^{57}Co is formed along with neutrons. The reaction may be written:

$$^{56}\text{Fe}_{26} + {}^{2}\text{H}_{1} \rightarrow {}^{57}\text{Co}_{27} + {}^{1}n_{0}. \qquad (1.4)$$

The ^{57}Fe formed as in eq. (1.3) is again in a nuclear excited state, and three gamma rays (14, 123, 137 keV) are emitted on decay to the stable ^{57}Fe ground state (Fig. 1.1). In addition to the gamma rays, X-rays are emitted

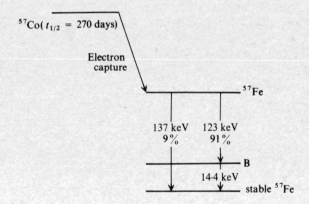

Fig. 1.1 The decay of ^{57}Co to stable ^{57}Fe. The 14·4 keV ray is the common Mössbauer photon.

as a consequence of the electron vacancy created after electron capture. Other electrons, denoted Auger electrons, may be ejected as a consequence of the rearrangement of the electrons in the atom.

Alpha emitters are confined, both in naturally occurring and artificially produced nuclides to the mass region above $A' \simeq 140$. Alpha particles are helium nuclei, and are denoted ^{4}He$_2$. An example of an alpha emitter is given by ^{241}Am.

$$^{241}\text{Am}_{95} \rightarrow {}^{237}\text{Np}_{93} + {}^{4}\text{He}_{2}. \qquad (1.5)$$

Mössbauer spectroscopy

Once again, the ^{237}Np emits a number of gamma rays on its way to the ground state, which in this case is also radioactive.

As noted several times above, the β and α decay processes often leave the resulting nucleus in an excited state. The de-excitation process usually involves the emission of high energy electromagnetic radiation—termed gamma (γ) radiation. Most often, the transition from excited to ground state proceeds through a number of intermediate excited states, and thus a number of gamma rays are emitted (e.g., Fig. 1.1).

Gamma emission is usually accompanied, or sometimes replaced, by a process called internal conversion. By the interaction between the excited nucleus and the extranuclear electrons, an electron may be emitted, with its energy E_e, equal to the difference between the nuclear transition energy, E_t, and the electron binding energy E_b

$$\text{i.e.} \quad E_e = E_t - E_b \tag{1.6}$$

1.2 Radioactive decay and the Heisenberg uncertainty principle

The decay of a radioactive substance via β or α processes, or the decay of a nuclear excited state via γ emission follows the exponential law $N = N_0 e^{-\lambda t}$ where N is the number of atoms in the initial unstable state at time t, N_0 is the number present when $t = 0$, and λ is a constant characteristic of the radioactive species. The rate of the radioactive decay is characteristically given by the half life, $t_{1/2}$, which is the time taken for the initial N_0 atoms to be reduced to $\frac{1}{2}N_0$. At $t = t_{1/2}$, it can be easily shown that

$$t_{1/2} = 0.693/\lambda \tag{1.7}$$

Half lives for these radioactive processes vary between about 10^{-10} seconds and 10^{15} to 10^{17} years, although these values probably represent the present limits of the sensitivity of radiometric methods. For example ^{241}Am has a half life of 458 years, while ^{57}Co has a half life of 270 days and ^{61}Cu has a half life of 3 hours. Nuclear excited states which emit gamma rays normally have much shorter half lives—usually in the range 10^{-6} to 10^{-10} seconds. For example, the first excited state of ^{57}Fe (B, Fig. 1.1) has a half life of $\sim 10^{-7}$ seconds, while one of the excited states of ^{61}Ni has a half life of $\sim 5 \times 10^{-9}$ seconds.

The emitted gamma rays are not monoenergetic (although as we will see shortly, their energy spread is usually very narrow), but have an energy distribution about a mean energy E_γ. This distribution is Lorentzian in shape (eq. 2.24), and is characterized by E_γ, the width at half height Γ_{ex}, and the area A. The minimum full width at half height Γ_H, which determines the sensitivity of the Mössbauer experiment, is determined by the mean life $\tau(=t_{1/2}/0.693)$ of the nuclear state and the Heisenberg uncertainty principle:

$$\Gamma_H \tau = h/2\pi = \hbar, \qquad (1.8)$$

where h = Planck's constant = 6.626×10^{-34} joule-sec. Remembering that 1 joule = 6.24×10^{18} eV (Appendix 1) and solving for Γ_H, we obtain:

$$\Gamma_H(\text{eV}) = 6.58 \times 10^{-16}/\tau \qquad (1.9\dagger)$$

or

$$\Gamma_H(\text{eV}) = 4.56 \times 10^{-16}/t_{1/2} \qquad (1.10)$$

Considering the excited state B of ^{57}Fe which has a $t_{1/2}$ of 97.7×10^{-9} seconds, Γ_H becomes 4.67×10^{-9} eV, or 0.097 mm s^{-1} as seen later in this chapter (Appendix 3). Thus, the line width is infinitesimal compared to E_γ, i.e., $\Gamma_H/E_\gamma = 3.24 \times 10^{-13}$ and the energy quanta are of unparalleled precision. By comparison, for a photon of 10 eV energy ($\sim 80\,000$ cm^{-1}) (Appendix 1) in the u.v. region, Γ_H can be of the order of 10^{-8} eV and $\Gamma_H/E_\gamma \sim 10^{-9}$, while the experimentally observed Γ_{ex}/E_γ is closer to 10^{-7} because of Döppler broadening (see later).

From eq. (1.10), it is apparent that as $t_{1/2}$ becomes smaller, Γ_H becomes larger; as $t_{1/2}$ becomes larger Γ_H becomes smaller. Thus for $t_{1/2} = 10^{-5}$ seconds, $\Gamma_H = 4.56 \times 10^{-11}$ eV, and for $t_{1/2} = 10^{-10}$ seconds, $\Gamma_H = 4.56 \times 10^{-6}$ eV. These two examples represent the limits of line widths for gamma transitions used in Mössbauer spectroscopy.

1.3 Nuclear properties

Electrons, protons, and neutrons have an intrinsic angular momentum about their own central axis of $\frac{1}{2}\hbar$. It is not surprising then that some nuclei possess angular momenta. This property is usually expressed by the spin, I, which is an integral or half-integral multiple of $\hbar$. Since the number of protons plus neutrons is given by A', and each proton or neutron can only add or subtract its angular momentum of $\frac{1}{2}\hbar$, even A' nuclei have integral spins (including zero) while odd A' nuclei have half-integral spins.

Quantum theory demands that the allowable nuclear spin states are quantized: that is, if a nucleus has a spin I, it can have component levels m_I which can only take the values, $m_I = I, I-1 \ldots -I$. For example, if a nucleus has a spin $I = \frac{1}{2}$, there are only two possible values of m_I, $+\frac{1}{2}$ and $-\frac{1}{2}$; if a nucleus has a spin $I = \frac{3}{2}$, there are four possible values of m_I, $+\frac{3}{2}, +\frac{1}{2}, -\frac{1}{2}$, and $-\frac{3}{2}$. In some situations, these states having different m_I will have the same energy (they are degenerate); whereas in other cases, extranuclear forces (magnetic and electric fields) interact with the nucleus to remove the degeneracy of the m_I levels.

† 1 eV = 1.6020×10^{-19} joules (Appendix 1).

Mössbauer spectroscopy

Nuclei with $I \neq 0$ give rise to a magnetic moment U_N given by the equation:

$$U_N = g_N \beta_N I \tag{1.11}$$

where: g_N is the nuclear Landé g factor

β_N, the nuclear Bohr magneton, $= \dfrac{e\hbar}{2mc}$

where e and m are the charge and mass of the proton, and c is the velocity of light.

If a steady magnetic field H is applied to a nuclear spin I, or the electronic environment imposes a permanent field H, there is an interaction between the field and the magnetic moment U_N such that $2I + 1$ energy levels are formed with energies given by

$$E_m = -g_N \beta_N H' m_I \tag{1.12}$$

where: H' is the magnetic field that the nucleus 'sees'—in general, different than the applied field H

and $m_I = I, I - 1 \ldots -I$ as above.

The degeneracy of the levels is completely removed and the separations between all the levels are identical ($= g_N \beta_N H'$). The magnitude of the splitting will be proportional to both the magnitude of g_N and H'.

In addition to a magnetic dipole moment, a nucleus of spin $I \geq 1$ has a quadrupole moment, Q, which may interact with local electric fields to split the degeneracy of the nuclear levels. A quadrupole moment may be thought of as arising from an elliptic charge distribution in the nucleus. Suppose that the charge distribution is somewhat elongated and is acted on by charges as in Fig 1.2, where ligand L_1 is more negative than ligand L_2.

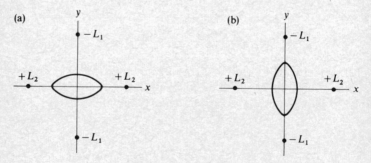

Fig. 1.2 Representation of a non-spherical nucleus in two orientations in the field of four charges. Configuration *b* is favourable energetically because the positive nuclear charge is closest to the negative ligand charge.

Figure 1.2b will correspond to lower energy, since it has the positive nuclear charge closest to the negative external charges. These two orientations correspond to two different sets of m_I values. For $I = \frac{3}{2}$, the nucleus can take up just two orientations corresponding to the states $m_I = \pm\frac{3}{2}$ and $m_I = \pm\frac{1}{2}$. The separation of these levels is called the quadrupole splitting (Q.S.) where:

$$\text{Q.S.} \propto qQ \qquad (1.13)$$

where q is proportional to the Z component of electric field gradient (see chapter 2, section 2).

For higher non-integral spins, quadrupole splitting results in $I + \frac{1}{2}$ levels; for integral spins, $2I + 1$ levels may be obtained in the most general case. The quadrupole interaction usually only partially removes the degeneracy of the nuclear levels, whereas the magnetic interaction completely removes the degeneracy.

Finally, in connection with the Mössbauer centre shift (chapter 2.1), the nuclear radius is important. Generally, the radius is determined from the cross-sectional area that the nucleus presents to high energy particles such as protons and alpha particles. The nuclear radii from these experiments are those distances from the centres of nuclei within which both coulombic (repulsive) and specific nuclear (attractive) forces act.

1.4 Döppler effect

The radioactive source used in Mössbauer experiments (e.g., ^{57}Co) emits gamma rays of various (different) energies, one of which is selected electronically to be the source photon. This photon, as seen earlier in this chapter, has a very narrow line width given by the Heisenberg uncertainty principle. In all forms of spectroscopy, the source photon energy has to be varied in some way. For example, in infrared spectroscopy, modulation of the source photon energy is accomplished by means of a grating. In Mössbauer spectroscopy, modulation of the source gamma ray energy is accomplished by the Döppler effect.

The Döppler effect is the phenomenon in which apparent sound or electromagnetic frequencies are altered by motion of the emitter relative to the observer. Döppler worked out his theory for sound waves in 1842, and then called attention to the phenomenon for the investigation of the nature and behaviour of light. Consider a photon source having an energy E_γ and a velocity v_0, moving directly towards the observer. The change in energy (to first order) of the photon ΔE_s, as seen by the observer is given by:

$$\Delta E_s = (v_0/c)E_\gamma \qquad (1.14)$$

If the source is moving at an angle θ from the line subtended by source and observer, then ΔE_s is given by $(v_0/c)E_\gamma \cos\theta$.

Mössbauer spectroscopy

In the Mössbauer experiment, the source of gamma rays (such as a ^{57}Co source) is mounted on a vibrator, and the desired range of photon energies is produced by vibrating the source. The changes in energy produced by moderate velocities are very small in comparison to E_γ. For example, consider the 14.4 keV ^{57}Fe gamma ray once again. For this isotope, a velocity of $1\,\mathrm{cm\,s^{-1}}$ corresponds to an energy (eq. 1.14) $\Delta E_s = 14.4 \times 10^3/3.00 \times 10^{10} = 4.80 \times 10^{-7}$ eV. Note that this is minuscule in comparison to E_γ, but two orders of magnitude larger than Γ_H derived earlier (4.67×10^{-9} eV). It is thus possible to produce large changes in the gamma ray energy (compared to Γ_H) with comparatively small velocities.

1.5 Resonant absorption and nuclear gamma resonance (the Mössbauer effect)

Resonant absorption is fundamental to most modern spectroscopic experiments. The basic equipment for most of these techniques includes a source of radiation whose energy can be varied, an absorber, and a detector. In the infrared region of the electromagnetic spectrum, (~ 100–$10\,000\,\mathrm{cm^{-1}}$), characteristic vibrations of a group of atoms can be excited resonantly if the incident radiation has an energy exactly equal to the difference in vibrational energy levels.

$$\Delta E_{\text{vibrational}} = h\nu_{\text{incident}} \tag{1.15}$$

We plot transmission (or absorption) versus energy of the source photon, and observe a peak or peaks when eq. (1.15) is satisfied (Fig. 1.3). From Fig. 1.3, it can be seen that for resonance to be observable, the observed line widths Γ_{ex} have to be smaller than the total possible source energy scan (ΔE_s), i.e., $\Gamma_{\text{ex}} < \Delta E_s$. In Fig. 1.3, the line widths (~ 5–$50\,\mathrm{cm^{-1}}$) are much smaller than the total energy scan, ΔE_s, of $\sim 1300\,\mathrm{cm^{-1}}$.

Returning now to nuclear gamma rays, we saw earlier that $\Gamma_H \ll \Delta E_s$. For ^{57}Fe, $\Gamma_H/\Delta E_s$ equalled approximately 10^{-2}. From this criterion, then,

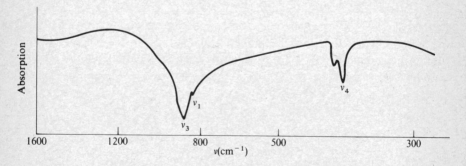

Fig. 1.3 Infrared spectrum of K_2CrO_4 (after K. Nakamoto, *Infrared Spectra of Inorganic and Coordination Compounds*, Wiley 1963).

there would appear to be no reason why nuclear gamma resonance should not be observable: i.e., we might expect that using a ^{57}Co source on a vibrator, an ^{57}Fe compound, and a suitable detector and counting equipment, we could plot transmission or absorption of gamma rays versus energy (specifically velocity) and observe a peak or peaks as in Fig. 1.3. However, many scientists, since Kuhn in 1929, tried to observe nuclear resonance, but experiments were rendered very difficult by several important differences between infrared resonance and nuclear gamma resonance. These differences will now be outlined, as they are fundamental to an understanding of Mössbauer's discovery in 1957.

Gamma rays are very energetic photons, having energies between 10 keV and 10 MeV. By comparison, infrared radiation has an energy of the order of 10^{-1} eV (Table 1.1). On emission of any photon, the emitting atom recoils

Table 1.1 Important energies (in eV) (in order of increasing values)

1. Free molecule recoil energy for infrared emitter	$\sim 10^{-12}$
2. Heisenberg natural line widths	10^{-10}–10^{-6}
3. Döppler shift for 14.4 keV gamma emitter moving at 1 cm s^{-1}	4.80×10^{-7}
4. 'Narrow' infrared line widths	$\sim 10^{-5}$
5. Free atom recoil energies for Mössbauer atom of $A' = 100$	10^{-3}–10^{-1}
6. Lattice vibrations—phonon energies	10^{-3}–10^{-1}
7. Döppler broadening (300 K for $A' = 100$)	10^{-2}–10^{-1}
8. Infrared radiation	10^{-2}–1
9. Mössbauer gamma energies	10^4–10^5
10. Gamma ray energies	10^4–10^7
11. 1 mass unit	9.31×10^8

(think of a bullet shot from a gun) and from conservation of momentum and energy, the effects of this recoil can be calculated. Consider a free atomic system of mass M with two energy levels E_e and E_g separated by a transition energy E_t (Fig. 1.4). If the system decays from E_e to E_g by emission of a photon of energy E_γ, conservation of momentum demands that the change in momentum of the emitting atom P_a be equal and opposite to the momentum of the gamma ray $P_\gamma (= E_\gamma/c)$. Squaring each momentum, and dividing both by $2M$, we have:

$$\frac{P_a^2}{2M} = \frac{-P_\gamma^2}{2M} = \frac{-E_\gamma^2}{2Mc^2} \tag{1.16}$$

Since $P_a = Mv$ (where v = recoil velocity), then $P_a^2/2M = E_R$, the kinetic energy of recoil. Thus

$$|E_R| = \frac{E_\gamma^2}{2Mc^2} \tag{1.17}$$

Mössbauer spectroscopy

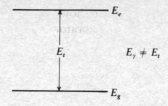

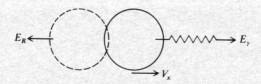

Fig. 1.4 Emission of a gamma ray of energy E_γ from an excited nuclear state of a nucleus having a velocity V_x. E_R is the recoil energy, and $E_t = E_e - E_g$.

E_R is always very small compared to the energy of an emitted photon. However, as we will see later, in contrast to infrared or u.v. emission for which $E_R \ll \Gamma_H$, for nuclear gamma emission $E_R \gg \Gamma_H$. This fact is of fundamental importance to the difficulties encountered in attempts to observe nuclear gamma resonance.

Consider now conservation of energy for the above free atomic system of mass M. We consider the problem in just one dimension without loss of generality. If the excited nucleus (energy E_e) has a velocity V_x at the moment before emission, then its total energy is $E_e + \frac{1}{2}MV_x^2$. A gamma ray of energy E_γ is then emitted in the x direction, the nucleus is de-excited to the state E_g, and the system will have a new velocity $(V_x + v)$ due to recoil (v can be negative). From conservation of energy:

$$E_e + \tfrac{1}{2}MV_x^2 = E_g + E_\gamma + \tfrac{1}{2}M(V_x + v)^2 \tag{1.18}$$

Rearranging, and noting that $E_t = E_e - E_g$, we have:

$$E_t - E_\gamma = \tfrac{1}{2}Mv^2 + MvV_x \tag{1.19}$$

$$= E_R + E_D \tag{1.20}$$

where: E_R, the recoil kinetic energy, can be expressed by eq. (1.17). (Note that it is independent of the initial velocity V_x of the system.)

and E_D, a velocity dependent term is denoted as a Döppler term.

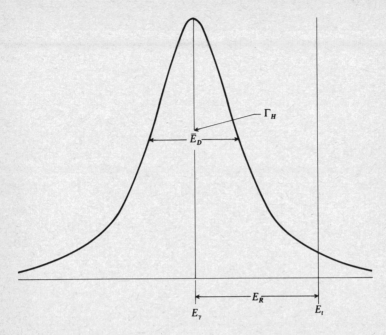

Fig. 1.5 Relationship of E_γ, E_t, E_R, $\bar{E}_D$, and Γ_H. Note that $E_t - E_\gamma = E_R$, and that E_R and $\bar{E}_D$ are orders of magnitude larger than Γ_H. On this scale, Γ_H is negligible and is represented by a line.

It is immediately apparent from eqs (1.19) and (1.20) that $E_t \neq E_\gamma$. E_γ is *smaller* than E_t by the recoil energy E_R and the Döppler term E_D (Fig. 1.5). E_R results in a shift of the photon energy from E_t to E_γ, but E_D, being dependent on the velocity of the emitting atom, results in a thermal broadening of the line (Fig. 1.5). To obtain a more convenient expression for $\bar{E}_D$, recall that for a random thermal motion of free atoms, the mean kinetic energy is given by $\frac{1}{2}M\bar{V}_x^2 \approx kT$, where $\bar{V}_x^2$ is the mean-square velocity of the atoms, k is Boltzmann's constant, and T is the absolute temperature. Rearranging the above expression, we have:

$$\bar{V}_x = (2kT/M)^{1/2} \quad (1.21)$$

and substituting $\bar{V}_x$ and $v = (2E_R/M)^{1/2}$ (eqs. 1.19, 1.20), we obtain

$$\bar{E}_D = Mv\bar{V}_x = 2(E_R kT)^{1/2} \quad (1.22)$$

What are the magnitudes of E_R and $\bar{E}_D$, and why do they seriously affect nuclear gamma resonance and not infrared resonance? A plot of approximate E_R and $\bar{E}_D$ values for an emitting mass number A' of 100 at 300 K is shown in Fig. 1.6. To calculate these energies, consider the emission of an 100 keV gamma ray by the above mass. Recalling, from Einstein's relationship

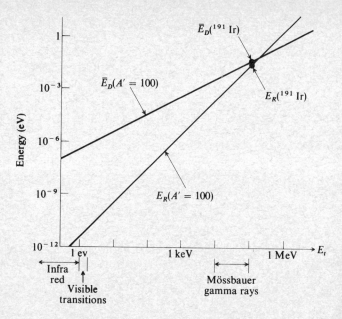

Fig. 1.6 Plot of $\bar{E}_D$ and E_R, for free atoms of $A' = 100$ at 300 K, as a function of the emitted photon energy, $\bar{E}_D$ and E_R for ^{191}Ir are noted (adapted from H. Frauenfelder, *The Mössbauer Effect*, W. A. Benjamin, 1962).

$E = mc^2$, that one mass unit corresponds to 931.5 MeV,† we can calculate (using eq. 1.17) that $E_R = 1 \times 10^{10}/(9.31 \times 10^8)(200) = 5.37 \times 10^{-2}$ eV. More generally:

$$E_R(\text{eV}) = \frac{5.37 \times 10^{-4} E_\gamma^2 \, (\text{keV})}{A'} \tag{1.23}$$

Note that these recoil energies for such gamma rays are much *larger* than the Heisenberg line widths calculated previously for gamma rays and $E_t \neq E_\gamma$ (Fig. 1.5). However, for photons of energy 1 eV (~ 8000 cm^{-1}), E_R is only $\sim 10^{-12}$ eV, and this is much *smaller* than Heisenberg line widths. Thus for u.v., visible, and i.r. radiation, $E_\gamma = E_t$ to a very good approximation.

Typical Döppler energies are also shown in Fig. 1.6 for free atoms at 300 K. These can be easily calculated using eq. (1.22) and taking $kT(300\,\text{K}) = 4.14 \times 10^{-21}$ joules $= 2.58 \times 10^{-2}$ eV. For example, if $E_\gamma = 100$ keV, $\bar{E}_D = 7.4 \times 10^{-2}$ eV. Note that this energy is comparable to E_R, but much larger than Γ_H.

† For 1 a.m.u., $m = 1.660 \times 10^{-27}$ K grams; $c = 2.998 \times 10^8$ m s^{-1}; $E = mc^2 = 1.492 \times 10^{-10}$ joules $= 931.5$ MeV (Appendix 1).

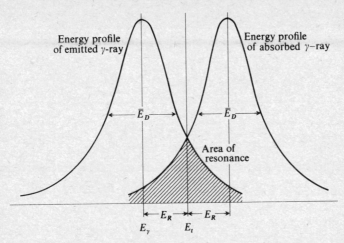

Fig. 1.7 Free atom gamma ray energy distributions for the ^{191}Ir 129 keV gamma ray. $E_R = 4.7 \times 10^{-2}$ eV, $\bar{E}_D = 7.0 = 10^{-2}$ eV, $\Gamma_H = 4.9 \times 10^{-6}$ eV. The area of resonance is shaded.

For ^{191}Ir, the above recoil effects give rise to the calculated energy profiles shown in Fig. 1.7. The emitted gamma rays have a distribution centred about $E_t - E_R$, while for resonant absorption to take place, the energy required by the absorbing atom (if the absorbing atom recoils) is centred about $E_t + E_R$. For ^{191}Ir, $\bar{E}_D > E_R$ at room temperature, and an appreciable amount of resonant absorption can be observed.

It is very difficult to make up for this recoil energy loss in order to increase the amount of resonant absorption. Since $2E_R \sim 10^{-1}$ eV, the Döppler velocity required for complete overlap of source and absorber energy profiles is (eq. 1.14): $v \simeq 0.1 \times 3 \times 10^{10}/1.3 \times 10^5 \simeq 2 \times 10^4$ cm s^{-1}. Such velocities are not readily achieved making it difficult to increase resonant absorption appreciably in this way for ^{191}Ir. It is also apparent that the line widths $\bar{E}_D (\sim 10^{-1}$ eV$)$ are much larger than a source energy scan ($\sim 10^{-5}$ eV) obtained by vibrating the source at velocities of the order of cm s^{-1}, and resonance is not easily observed by a conventional plot of absorption versus readily obtainable velocities.

In contrast to the above situation, the recoil energy E_R ($\sim 10^{-12}$ eV) for infrared emission is negligible compared to both $\bar{E}_D$ ($\sim 10^{-6}$ eV) and natural line widths Γ_H. Emission and absorption energy profiles overlap completely, and resonance is easily achieved. The Döppler broadening (and instrumental effects) determine the line width rather than the Heisenberg line width Γ_H.

It was Mössbauer's great discovery that for some low energy gamma rays, E_R and $\bar{E}_D$ became negligible such that source and absorber energy profiles overlapped completely and line widths approaching Γ_H were observed. When studying nuclear resonant absorption using ^{191}Ir, he observed that the resonant effect *increased* on cooling the sample. He expected the effect to *decrease*, because $\bar{E}_D$ decreases as T decreases (eq. 1.22) leading to less overlap of source and absorber energy profiles. With arguments along the following lines, Mössbauer postulated that a significant fraction of the gamma rays were emitted without recoil.

Consider, first, an excited nucleus *rigidly* held in a solid. On gamma emission, the emitting atom cannot be ejected from its lattice position, since E_R is much less than chemical binding energies (Table 1.1). The recoiling mass is now the mass of the whole crystal which may have $\sim 10^{17}$ atoms. The recoiling mass in eq. (1.23) is enormous, and E_R now becomes $\sim 10^{-19}$ eV for a 100 keV gamma ray and $A' = 100$. E_R then is much smaller than Γ_H. Similarly, considering $\bar{E}_D$, the atom in this approximation cannot undergo random thermal motion (since it is rigidly held) and $\bar{E}_D$ becomes negligible compared to Γ_H. Thus, for a rigidly held atom, $E_t \simeq E_\gamma$, source and absorber energy profiles will completely overlap and will both have widths approaching Γ_H.

However, the emitting atom is not rigidly bound, but is usually free to vibrate. The recoil energy could then be transferred to exciting a lattice vibration, whose energies (Table 1.1) are comparable to E_R. But the lattice, like nuclei and electrons, is a quantized system which cannot be excited in an arbitrary fashion. On a simple Einstein model of a solid, an energy of $\pm h\nu$, $\pm 2h\nu$, $\pm 3h\nu$ etc., is required to excite the lattice. The recoil energy E_R must be equal to or greater than the energy required to excite the lattice to its lowest excited level, i.e., $E_R \geq h\nu$. If E_R is less than this (and for $E_\gamma = 10$ keV, $E_R \sim 10^{-4}$ eV), then the lattice will not be excited, the emitting atom effectively does not recoil, and the whole crystal mass takes up the recoil. Again E_R and $\bar{E}_D$ become minute, $E_t \simeq E_\gamma$, and resonance is easily observable.

Because E_R and lattice excitation energies are of comparable magnitude, only a certain fraction of the emissions and absorptions take place without recoil, and this fraction will vary from solid to solid, as well as decreasing as E_γ (and thus E_R) increase. This fraction f is known as the Mössbauer fraction, and can be expressed as:

$$f = \exp\left[\frac{-4\pi^2 \langle X^2 \rangle}{\lambda^2}\right] \tag{1.24}$$

where:

λ is the wavelength of the gamma photon, and $\langle X^2 \rangle$ is the mean square vibrational amplitude of the emitting (or absorbing) nucleus in the solid.

The larger λ becomes (the smaller E_γ), the larger f becomes; while as $\langle X^2 \rangle$ increases, f decreases. For a finite f value, E_γ should be under 150 keV. Since $\langle X^2 \rangle$ has to be bounded in order that f does not vanish, the Mössbauer effect cannot normally be observed in a liquid. It is conceivable that $\langle X^2 \rangle$ will remain small enough in a viscous liquid to allow detection of a recoil-free event. f values are usually large for ionic type solids such as silicate minerals and much lower for organometallic compounds.

If we now take a suitable Mössbauer nuclide such as ^{57}Fe in stainless steel as an absorber and ^{57}Co in stainless steel as a source, vibrate the source over a range of velocities and count the number of gamma rays transmitted through the absorber at each velocity, the one line spectrum in Fig. 1.8a results. The one peak at zero velocity reflects the fact that both source and

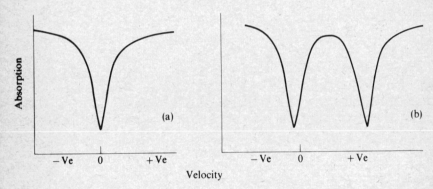

Fig. 1.8 Mössbauer spectra of *a*, a stainless steel absorber using a ^{57}Co in stainless steel source and *b* a general absorber using the above source.

absorber energy levels are identical—as one would expect for ^{57}Fe atoms in identical environments. However, if one keeps the same single line source and changes the absorber, one or more peaks will be observed, none of which in general will be at zero velocity (Fig. 1.8b). The nuclear energy levels are thus sensitive to the extranuclear environment. The next chapter outlines the theory used to interpret these changes in peak positions.

Problems

1. For the following $t_{1/2}$ values for the excited nuclear states, calculate the widths at half height (Γ_H) in electron volts and cm^{-1}: ^{67}Zn $(t_{1/2} = 9.3 \times 10^{-6}$ sec), ^{119}Sn $(t_{1/2} = 1.9 \times 10^{-8}$ sec), ^{191}Ir $(t_{1/2} = 9.4 \times 10^{-11}$ sec).

Mössbauer spectroscopy

Take $h = 6.626 \times 10^{-34}$ joule-sec, 1 joule $= 6.24 \times 10^{18}$ eV, and 1 eV $= 8066$ cm^{-1}.

2. For the above three nuclei, calculate the free atom recoil energies, and the free Döppler widths (at 300 K) in the gas phase. Draw the energy profile for emission and absorption (see Fig. 1.7). $E_\gamma(\text{Zn}) = 93.3$ keV; $E_\gamma(\text{Sn}) = 23.9$ keV; $E_\gamma(\text{Ir}) = 129.5$ keV.
3. If recoil does occur for these three nuclei, what velocity would have to be provided to the source for complete overlap of source and absorber lines (assuming that source and absorber lines remain at the same width)? Comment on the difficulty in observing a large fraction of resonant absorption in these isotopes.
4. By contrast, calculate the free molecule recoil energy for an infrared photon of energy 807 cm^{-1} absorbed by IrO_2. Compare this with typical infrared linewidths.
5. Convert the natural linewidths (in eV) in problem 1 to mm s^{-1}. Assuming that recoilless emission and absorption is observed for these three isotopes (and the natural line widths are observed), what range of velocities (very approximately) would have to be scanned using a 400 channel analyzer in a Mössbauer experiment? Give possible reasons why ^{67}Zn and ^{191}Ir may not be useful Mössbauer isotopes from this point of view.
6. The mean square vibrational amplitude $\langle X^2 \rangle$ of ^{57}Fe in a chemical compound is 7.1×10^{-4} Å^2. Calculate the f value for the 14.4 keV emission for this iron atom.
7. The Mössbauer effect has been observed for two gamma photons in ^{57}Fe, the above 14.4 keV photon and a 136.3 keV photon. Using the same value for $\langle X^2 \rangle$ as in the previous problem, why is the 136.3 keV photon not used more often for chemical work?

Bibliography

1.1 G. Friedlander, J. W. Kennedy and J. M. Miller, *Nuclear and Radiochemistry*, 2nd edition, Wiley and Sons, 1966.
1.2 G. K. Wertheim, *Mössbauer Effect: Principles and Applications*, Academic Press, 1964.
1.3 H. Frauenfelder, *The Mössbauer Effect*, W. A. Benjamin, New York, 1962.
1.4 N. N. Greenwood, Mössbauer Spectroscopy, in *Physical Chemistry, an Advanced Treatise*, vol. 4, p. 633, Academic Press, 1970, Eds. H. Eyring, D. Henderson, and W. Jost.

2. Mössbauer parameters and theory

Figure 1.8 indicated that the peak positions in a Mössbauer spectrum are sensitive to the extranuclear environment, such that different compounds give different spectra. The differences in spectra can be attributed to the so-called hyperfine interactions—the interactions between the nuclear charge distribution and the extranuclear electric and magnetic fields. These hyperfine interactions give rise to the isomer shift (I.S.), the quadrupole splitting (Q.S.) and the magnetic Zeeman splitting. We will be concerned mainly with the first two parameters, and how they can be related to the electronic and ligand structure of compounds and minerals.

In addition, other spectral parameters such as peak shapes, widths, and areas will be discussed at the end of this chapter. These parameters are of special importance for the detailed interpretation of the mineral spectra.

2.1 The isomer shift

The isomer shift results from the electrostatic interaction between the charge distribution of the nucleus and those electrons which have a finite probability of being found in the region of the nucleus. The above interaction does not lead to a splitting of the nuclear energy levels, but rather results in a slight shift of the Mössbauer energy levels in a compound relative to those in the free atom. The shift will in general be different in source and absorber (Fig. 2.1a), and thus the energy of the source gamma ray $^sE_\gamma$, and the energy required for resonant absorption, $^aE_\gamma$, will be different by $\sim 10^{-9}$ eV. A Döppler velocity will have to be supplied to the source or absorber to observe resonance: $^sE_\gamma \pm (v/c)^sE_\gamma = {}^aE_\gamma$. In this case, a one line spectrum of the type shown in Fig. 2.1b results.

Before discussing the origin of the isomer shift, it is important to realize that such a shift of energy levels can also arise from the second order Döppler (S.O.D.) shift which arises from the thermal motion of the Mössbauer atoms. The centre or chemical shift (C.S.) observed in a Mössbauer spectrum is a resultant of both the isomer shift and second order Döppler shift, but the second order Döppler shift is usually much smaller than the isomer shift,

Mössbauer spectroscopy

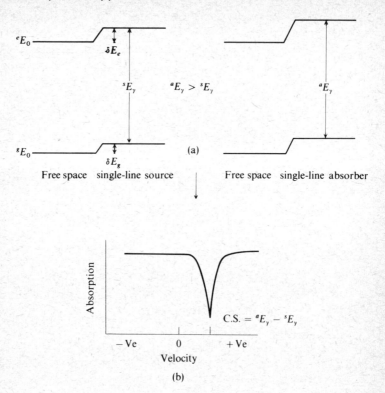

Fig. 2.1 Nuclear energy levels and the isomer shift. (a) Source and absorber nuclear energy levels, (b) resultant Mössbauer spectrum.

and the variations in the second order Döppler shift from compound to compound are very small. The second order Döppler shift has usually been neglected, but when discussing results in later chapters, the term *centre shift* is used when referring to the experimentally derived parameter. Other terms such as chemical shift and chemical isomer shift have also been used and refer to the same parameter.

The isomer shift can be computed classically by considering the effect of the overlap of s-electron density with the nuclear charge density. Only s electrons of the hydrogen-like orbitals have a finite probability of overlapping with the nuclear charge density, and thus of interacting with it. As we will see shortly, however, the s-electron density at the nucleus is often sensitive to the p- or d-electron density.

The nucleus is assumed to be a uniformly charged sphere of radius R, and the s-electron density at the nucleus, $[\Psi(0)_s]^2$, is assumed to be a constant over the nuclear dimensions. One computes the difference between the electrostatic interaction of a point nucleus with $[\Psi(0)_s]^2$, and the interaction of a nucleus having a radius R with $[\Psi(0)_s]^2$. The difference in energy is given by:

$$\delta E = K[\Psi(0)_s]^2 R^2 \qquad (2.1)$$

where K is a nuclear constant. Since R is generally different for ground and excited nuclear states, δE will be different for both (Fig. 2.1a), and:

$$\delta E_e - \delta E_g = K[\Psi(0)_s]^2 (R_e^2 - R_g^2) \qquad (2.2)$$

where the subscripts e and g refer to the excited and ground nuclear states respectively. The R values are nuclear constants, but the $[\Psi(0)_s]^2$ values will vary from compound to compound. The above energy difference becomes measurable in a Mössbauer experiment by comparing the nuclear transition energy in a source ($^sE_\gamma$) with that in an absorber ($^aE_\gamma$). The isomer shift is the Döppler velocity which is provided to the source to observe resonance, and is given by the difference of eqs. (2.2) for source and absorber, i.e.,

$$\text{I.S.} = K(R_e^2 - R_g^2)\{[\Psi(0)_s]_a^2 - [\Psi(0)_s]_s^2\} \qquad (2.3)$$

where the subscripts a and s refer to the absorber and source respectively. Usually, the source is a standard material e.g., ^{57}Co in Pd for Fe Mössbauer spectra, or BaSnO$_3$ for Sn Mössbauer spectra. Since the change in radius $R_e - R_g$ is very small, the isomer shift can then be written in its usual form:

$$\text{I.S.} = 2KR^2 \frac{\delta R}{R}\{[\Psi(0)_s]_a^2 - C\} \qquad (2.4)$$

where: $\delta R = R_e - R_g$ and C is a constant characteristic of the source used. Thus the isomer shift depends on a nuclear factor δR and an extranuclear factor $[\Psi(0)_s]^2$. For a given nucleus, δR is a constant, so that the isomer shift is directly proportional to the s-electron density at the nucleus. When δR is positive (as for ^{119}Sn), an increase in s-electron density at the absorber nucleus results in a more positive isomer shift; when δR is negative (as for ^{57}Fe), an increase in s-electron density at the absorber nucleus results in a more negative isomer shift. Of course, the larger the magnitude of $\delta R/R$, the larger the change in isomer shift for a given change in s-electron density. Some values for $\delta R/R$ are given in Table 2.7, but for most isotopes, these values are not well known. These uncertainties make it difficult to calculate s-electron densities from isomer shifts.

Although changes in isomer shift are due to variations in s-electron density, differences in isomer shifts are observed on addition or removal of p or d electrons which do not themselves interact with the nuclear charge density. Hartree–Fock calculations for different d^n configurations by Watson (Table 2.1) show that a decrease in the number of d electrons causes a marked increase in the total s-electron density at the iron nucleus. This is due almost

Mössbauer spectroscopy

Table 2.1 Total s-electron densities at the iron nucleus for different d configurations

d Electron configuration	Electron density at the iron nucleus (atomic units)
d^8	11 878·6
d^7	11 879·1
d^6	11 879·5
d^5	11 881·3
d^4	11 885·2
d^3	12 392·0

(See R. H. Golding, *Applied Wave Mechanics*, Van Nostrand (1970), Chapter 9.)

entirely to changes in the 3s density at the nucleus. This trend arises indirectly via the 3s electrons which spend a fraction of their time further from the nucleus than the 3d electrons. Figure 2.2, which shows the radial distribution of 3s, 3p, and 3d electrons in a ferric ion, shows that the 3d electrons do have a finite probability of lying within the 3s electrons, causing the 3s electrons to expand and thus reduce the s-charge density at the nucleus. Remembering that $\delta R/R$ is negative for ^{57}Fe, one would expect that a d^6 ion (Fe^{2+}) would have an appreciably larger isomer shift than a d^5 ion (Fe^{3+}).

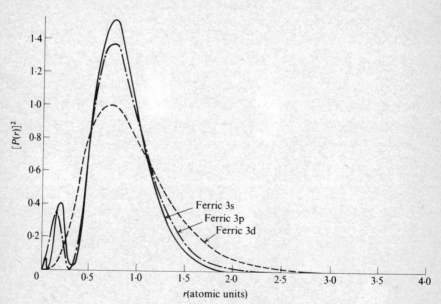

Fig. 2.2 Hartree–Fock radial charge distribution $[P(r)]^2$ for the 3s, 3p and 3d orbitals for the Fe^{3+} ion (R. M. Golding, *Applied Wave Mechanics*, Van Nostrand, 1969, p. 400).

In a molecule, the picture becomes more complex because both the s- and d-electron densities will be modified by covalent bonding. In ^{57}Fe, the two important bonding interactions (see chapter 5) contribute to a $4s$ population, and a change in the d orbital population from the free-ion value. Any increase in the $4s$ population decreases the isomer shift, and can be calculated using the Fermi–Segré–Goudsmit formula:

$$[\Psi(0)_{ns}]^2 = \frac{Z_i Z_o^2}{\pi a_o^3 n_o^3}\left(1 - \frac{d\sigma}{dn}\right) \quad (2.5)$$

where n is the principal quantum number of the s state, Z_i is the internal effective nuclear charge, Z_o is the external charge acting on the valence electrons, n_o and σ are the effective quantum number and quantum defect respectively of the nth state. The effect of a change of d-electron populations is more difficult to calculate. Several workers have calculated isomer shifts for $3d^n 4s^x$ configurations for ^{57}Fe, but for a variety of reasons these calculations are not entirely satisfactory (ref. 2.1). For our purposes, it is usually extremely useful just to obtain *relative* changes in valency orbital populations from *relative* isomer shifts, keeping in mind that an increase in $4s$ density decreases the isomer shift, while an increase in $3d$ density increases the isomer shift.

For ^{119}Sn, the sign of $\delta R/R$ was in dispute for some time. However, this is now known to be positive, and the isomer shift will thus increase with an increase in s-electron density but decrease with an increase in p-electron density. Lees and Flinn (ref. 2.2) have derived an equation for the isomer shift relative to Mg_2Sn as a function of the number of $5s$ and $5p$ electrons (designated n_s and n_p respectively):

$$\text{I.S.} = -2.36 + 3.01 n_s - 0.20 n_s^2 - 0.17 n_s n_p \ (\text{mm s}^{-1}) \quad (2.6)$$

Although the accuracy of this equation is debatable, it does indicate that the isomer shift should be much more sensitive to the s-electron density than the p-electron density. It would be expected that the isomer shifts for Sn species would increase in the following order: ionic Sn^{IV} (electronic configuration $4d^{10}$) < covalent Sn $(5s5p^3)$ < Sn^{II} $(5s^{2-x}5p^x)$.

The above discussion indicates that if the sign of $\delta R/R$ is known, the Mössbauer isomer shift provides first, a very useful, if at the present time only qualitative, method of examining the covalent character of a bond or bonds involving a Mössbauer isotope; and second, a potential method for determining the valency and oxidation state of the Mössbauer atom.

2.2 Quadrupole splitting

The expression for the isomer shift (eq. 2.4) was derived assuming the nucleus to be spherical and to have a uniform charge density. If these conditions

Mössbauer spectroscopy

are relaxed, and if $I > \frac{1}{2}$, the interaction of non-cubic extranuclear electric fields with the nuclear charge density results in a splitting of the nuclear energy levels. For example, for ^{57}Fe and ^{119}Sn, $I_e = \frac{3}{2}$ and $I_g = \frac{1}{2}$, and the $I = \frac{3}{2}$ level splits into two ($m_I = \pm\frac{3}{2}, \pm\frac{1}{2}$) while the $I = \frac{1}{2}$ levels remain degenerate (Fig. 2.3a). Both the possible transitions are allowed, and a characteristic two line spectrum is obtained (Fig. 2.3b). The separation of the peaks is the quadrupole splitting (Q.S.) and the centroid of the two peaks relative to the source is the centre shift (C.S.). In Fig. 2.3b, the quadrupole

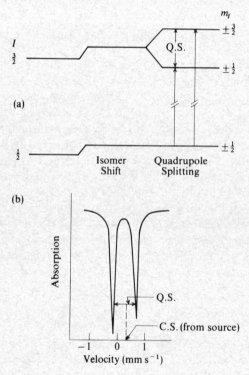

Fig. 2.3 Nuclear energy levels and the quadrupole splitting. (a) Absorber energy levels: excited level ($I = \frac{3}{2}$) split into two by quadrupole interaction, (b) resultant Mössbauer spectrum.

splitting is about 0.9 mm s^{-1} and the centre shift about 0.3 mm s^{-1}. For half-integral nuclear spins, the quadrupole interaction results in $I + \frac{1}{2}$ levels for spin I. For integral nuclear spins, the degeneracy of the nuclear levels may be completely removed by the quadrupole interaction to give $2I + 1$ levels. If both ground and excited nuclear states have large I, then more complex Mössbauer spectra can be observed. The example of ^{129}I with I_g and $I_e = \frac{7}{2}$ and $\frac{5}{2}$ respectively will be considered later in this chapter.

The magnitude of the quadrupole splitting is proportional to the Z component of the electric field gradient (EFG) tensor which interacts with the quadrupole moment of the nucleus. For the $I = \frac{3}{2}$ case (e.g., ^{57}Fe and ^{119}Sn), the quadrupole splitting can be expressed as:

$$\text{Q.S.} = \tfrac{1}{2}e^2qQ(1 + \eta^2/3)^{1/2} \qquad (2.7)$$

where: Q is the quadrupole moment of the nucleus

$eq = V_{ZZ} = $ -the Z component of the EFG.

$e = $ protonic charge $= 4.80 \times 10^{-10}$ esu $= 1.602 \times 10^{-19}$ coulomb.

$\eta = $ the asymmetry parameter $= (V_{XX} - V_{YY})/V_{ZZ}$.

Either q or V_{ZZ} are normally referred to as the field gradient, and for convenience, q will be referred to as such in the remainder of this book.

Ideally, we would like to obtain three pieces of information from a measurement of the quadrupole splitting (eq. 2.7): the magnitude of q and η, and the sign of the quadrupole splitting. If the $\pm\frac{3}{2}$ state is at high energy (as in Fig. 2.3), the sign of the quadrupole splitting is positive. For Mössbauer nuclei such as ^{57}Fe and ^{119}Sn having nuclear spins $\frac{1}{2}$ and $\frac{3}{2}$, a two line spectrum is observed and the quadrupole splitting is easily measured. But, by the nature of eq. (2.7), it is obvious that both q and η cannot be calculated from a measurement of the quadrupole splitting. Also, the sign of the quadrupole splitting cannot be determined (generally) from a powder spectrum since the $\pm\frac{3}{2}$ and $\pm\frac{1}{2}$ lines are of very similar intensity and cannot be distinguished. Techniques for measuring the sign of Q.S., and estimating η, for ^{57}Fe, ^{119}Sn, and isotopes having $I > \frac{3}{2}$, will be discussed later in this chapter.

The EFG tensor. What is the EFG tensor and how is it related to q and η in eq. (2.7), and to the electrical asymmetry about the Mössbauer atom? The EFG tensor has nine components which arise in the following way. The electric field at the Mössbauer nucleus is the negative gradient of the potential, V:

$$E = -\nabla V = -(\mathbf{i}V_x + \mathbf{j}V_y + \mathbf{k}V_z)$$

where:

$$V_x = \frac{\partial V}{\partial x},\ V_y = \frac{\partial V}{\partial y},\ V_z = \frac{\partial V}{\partial z}$$

As is perhaps evident, the EFG is the gradient of the electric field.

$$\text{EFG} = \nabla E = -\begin{bmatrix} V_{xx} & V_{xy} & V_{xz} \\ V_{yx} & V_{yy} & V_{yz} \\ V_{zx} & V_{zy} & V_{zz} \end{bmatrix} \qquad (2.8)$$

Mössbauer spectroscopy

where:

$$V_{xx} = \frac{\partial^2 V}{\partial x^2}, \quad V_{xy} = \frac{\partial^2 V}{\partial x \partial y}, \quad V_{xz} = \frac{\partial^2 V}{\partial x \partial z}, \text{ etc.}$$

If we assume that the EFG is set up by point charges Z_i, the contribution of one point charge to each component of the EFG tensor is given by the

Table 2.2 Components of the electric field gradient tensor for a charge of Z electronic units

$V_{xx} = Zer^{-3}(3\sin^2\theta\cos^2\phi - 1)$	$V_{xy} = V_{yx} = Zer^{-3}(3\sin^2\theta\sin\phi\cos\phi)$
$V_{yy} = Zer^{-3}(3\sin^2\theta\sin^2\phi - 1)$	$V_{xz} = V_{zx} = Zer^{-3}(3\sin\theta\cos\theta\cos\phi)$
$V_{zz} = Zer^{-3}(3\cos^2\theta - 1)$	$V_{yz} = V_{zy} = Zer^{-3}(3\sin\theta\cos\theta\sin\phi)$

(See Fig. 2.4 for definition of polar coordinates.)

expressions in Table 2.2. The polar coordinates are defined as usual (Fig. 2.4). The total contribution to each EFG component is obtained by simply summing the individual contributions. The above tensor can be reduced to diagonal form if the coordinate axes are properly chosen, so that the EFG can be completely specified by the three components, V_{XX}, V_{YY} and V_{ZZ}.† However, even these three are not independent, since

$$V_{XX} + V_{YY} + V_{ZZ} = 0. \quad (2.9)$$

Remembering that $\cos^2 x + \sin^2 x = 1 (x = \theta, \phi)$, it is evident from the expressions in Table 2.2 that eq. (2.9) is obeyed. There are, then, only two

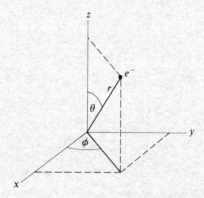

Fig. 2.4 Polar co-ordinates of an electron in free space.

† The principal EFG components are denoted by V_{XX}, V_{YY} and V_{ZZ} — i.e. with capital subscripts.

Mössbauer parameters and theory

independent parameters, and these are normally chosen to be V_{ZZ} and η, where $\eta = (V_{XX} - V_{YY})/V_{ZZ}$. The EFG axes are chosen such that the off-diagonal components are zero (if possible), and that $|V_{ZZ}| \geq |V_{YY}| \geq |V_{XX}|$, which constrains η to have values between 0 and 1. If the off-diagonal elements are non zero, then the tensor must be diagonalized before the diagonal components are chosen as above.

Many of the properties of the EFG tensor can be deduced from the symmetry properties of the molecule or crystal. Often, the Z EFG axis coincides with the highest symmetry axis of the molecule or crystal. Neglecting intermolecular contributions to the EFG tensor, and considering just point charge contributions from ligands bound to the Mössbauer atom, the Z EFG axis in trans-MA_2B_4(M = Mössbauer atom) lies along the four fold axis (Fig. 2.5a). To confirm that this satisfies the criteria given above, let

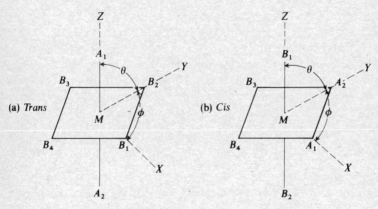

Fig. 2.5 EFG axes for *trans-* and *cis-*MA_2B_4.

$Z_A/r_A^3 = [A]$ and $Z_B/r_B^3 = [B]$. The contribution of each ligand to each component of the EFG tensor is given in Table 2.3. Summing these contributions, we find that:

$$V_{XX} = V_{YY} = (-2[A] + 2[B])e \qquad (2.10a)$$

$$V_{ZZ} = (4[A] - 4[B])e \qquad (2.10b)$$

$$V_{xz} = V_{yz} = V_{xy} = 0 \qquad (2.10c)$$

Thus

$$\eta \equiv 0, \quad V_{XX} + V_{YY} + V_{ZZ} = 0$$

and

$$|V_{ZZ}| > |V_{YY}| = |V_{XX}|.$$

Mössbauer spectroscopy

Table 2.3 Individual point charge contributions to the EFG tensor in *trans*- and *cis*-MA_2B_4

trans-MA_2B_4

Ligand (Fig. 2.5)	θ	ϕ	$\sin\theta$	$\cos\theta$	$\sin\phi$	$\cos\phi$	V_{XX}/e	V_{YY}/e	V_{ZZ}/e
A_1	0	0	0	1	0	1	$-[A]$	$-[A]$	$+2[A]$
A_2	180	0	0	-1	0	1	$-[A]$	$-[A]$	$+2[A]$
B_1	90	0	1	0	0	1	$+2[B]$	$-[B]$	$-[B]$
B_2	90	90	1	0	1	0	$-[B]$	$+2[B]$	$-[B]$
B_3	90	180	1	0	0	-1	$+2[B]$	$-[B]$	$-[B]$
B_4	90	270	1	0	-1	0	$-[B]$	$+2[B]$	$-[B]$

cis-MA_2B_4

Ligand	θ	ϕ	$\sin\theta$	$\cos\theta$	$\sin\phi$	$\cos\phi$	V_{XX}/e	V_{YY}/e	V_{ZZ}/e
A_1	90	0	1	0	0	1	$+2[A]$	$-[A]$	$-[A]$
A_2	90	90	1	0	1	0	$-[A]$	$+2[A]$	$-[A]$
B_1	0	0	0	1	0	1	$-[B]$	$-[B]$	$+2[B]$
B_2	180	0	0	-1	0	1	$-[B]$	$-[B]$	$+2[B]$
B_3	90	180	1	0	0	-1	$+2[B]$	$-[B]$	$-[B]$
B_4	90	270	1	0	-1	0	$-[B]$	$+2[B]$	$-[B]$

However, in other cases of interest, (e.g. *cis*-MA_2B_4, Fig. 2.5b), the Z EFG axis does not coincide with the highest symmetry axis of the molecule, in this case a two-fold axis. The choice given in the figure gives all off-diagonal elements as zero initially, whereas the choice of the Z axis coincident with the two-fold axis gives off-diagonal terms. These can be eliminated on diagonalizing the tensor but it is convenient to have no off-diagonal elements to begin with. Summing the contributions in Table 2.3, we obtain

$$V_{XX} = V_{YY} = ([A] - [B])e \tag{2.11a}$$

$$V_{ZZ} = (-2[A] + 2[B])e \tag{2.11b}$$

Again

$$\eta = 0, \text{ and } |V_{ZZ}| > |V_{YY}| = |V_{XX}|.$$

We will see in chapter 6 (Tables 6.13, #14) that the Z EFG axis, and the X and Y axes, do not always coincide with metal-ligand bond directions. In these cases, it is usually necessary to choose an arbitrary set of axes, work out the EFG components, diagonalize the tensor, and then choose $|V_{ZZ}| \geqslant |V_{YY}| \geqslant |V_{XX}|$.

So far, we have discussed the EFG tensor components just considering point charge contributions. However, electrons about the Mössbauer atom, if the electronic configuration has a symmetry lower than cubic, also contribute to the EFG tensor, and it is convenient to divide the field gradient into two contributions. For simplicity, we assume that $\eta = 0$, and express q as: (ref. 6.5)

$$q = (1 - \gamma_\infty)q_{lattice} + (1 - R)q_{valence} \qquad (2.12)$$

where: (a) R and γ_∞ are the Sternheimer antishielding factors,
(b) $q_{lattice}$, the contribution from the external ligand charges Z_i equals V_{zz}/e, where V_{zz} is given in Table 2.2, and
(c) $q_{valence}$, the contribution from the valence electrons equals

$$- \sum_{\substack{\text{Valence}\\\text{electrons}}} \langle 3\cos^2\theta - 1\rangle\langle r^{-3}\rangle \qquad (2.13)$$

Table 2.4 gives the contributions to $q_{valence}$ for p and d electrons, where $+\frac{4}{7}$ is $\langle 3\cos^2\theta - 1\rangle$ for d_{z^2}, etc. While the inner, non-valence filled shells have cubic symmetry and do not contribute directly to the field gradient, the presence of a $q_{lattice}$ or $q_{valence}$ will polarize and distort the inner electrons.

Table 2.4 Values of $q_{valence}$ for p and d electrons

Wave function	$V_{zz}/e = q_{valence}$†	Wave function	$V_{zz}/e = q_{valence}$
p_x	$+\frac{2}{5}\langle r^{-3}\rangle_p$	d_{z^2}	$-\frac{4}{7}\langle r^{-3}\rangle_d$
p_y	$+\frac{2}{5}\langle r^{-3}\rangle_p$	$d_{x^2-y^2}$	$+\frac{4}{7}\langle r^{-3}\rangle_d$
p_z	$-\frac{4}{5}\langle r^{-3}\rangle_p$	d_{xy}	$+\frac{4}{7}\langle r^{-3}\rangle_d$
		d_{xz}	$-\frac{2}{7}\langle r^{-3}\rangle_d$
		d_{yz}	$-\frac{2}{7}\langle r^{-3}\rangle_d$

† $\langle r^{-3}\rangle$ is the average of r^{-3}.

This polarization will magnify any $q_{valence}$ or $q_{lattice}$. The Sternheimer terms take this polarization into account. γ_∞ is often quite large (values of -10 are not uncommon), while R is thought to be much less than one in many cases (~ 0.2 to 0.3 for Fe and Sn). The Sternheimer factors therefore do not affect the signs of $q_{lattice}$ and $q_{valence}$.

The $q_{lattice}$ term can in principle be calculated using the equations in Table 2.2 if the crystal structure is known. However, there are several difficulties in this calculation. First, it is difficult to assign charges to different atoms in the structure, and the lattice summation assumes that the atoms are point charges, which is usually not valid; second, the lattice sum may converge very slowly with increasing r; third, $q_{lattice}$ is very sensitive to the position of the ions, and small errors in their positions could give appreciable differences in $q_{lattice}$. Even if q can be calculated accurately, the value of Q is not accurately known for many isotopes, and it is difficult then to compare calculated and observed quadrupole splittings. For example, for ^{57}Fe, estimates of Q vary from 0·15 to 0·4 barns.

Direct calculation of $q_{valence}$ can, in principle, be made using eq. (2.13) and Table 2.4. Such calculations would require a knowledge of the orbital populations of the valence orbitals for the Mössbauer atom. These are generally not known, and there are great difficulties in obtaining accurate estimates.

Because of these difficulties, more empirical approaches, such as the partial quadrupole splitting and ligand field treatments outlined in chapter 6, are often more useful for rationalizing quadrupole splittings in related series of compounds.

The sign of q. Both $q_{valence}$ and $q_{lattice}$ (and thus q) will be zero if the electron and ligand charge distributions respectively, have cubic symmetry. Both of these situations can be easily demonstrated. Consider a perfect octahedron of point charges, which are taken to represent six ligands. Each one is at a distance r from the nucleus.

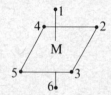

Using the equation for V_{zz} (Table 2.2), it is apparent that $(3\cos^2\theta - 1)$ is 2 for ligands 1 and 6, and -1 for ligands 2, 3, 4, and 5. Summing the six contributions gives $q_{lattice} = 0$. If we now compress ligands 1 and 6, such that $r_i(i = 1, 6) < r_j(j = 2, 3, 4, 5)$, then $q_{lattice}$ becomes negative; if, on the other hand $r_i > r_j$, then $q_{lattice}$ becomes positive. Thus concentrating the negative charge along the Z axis relative to X and Y will give rise to a negative $q_{lattice}$.

Using eq. (2.13) and Table 2.4, the $q_{valence}$ contribution from a p electron imbalance can be written as:

$$q_{valence} = K_p[-N_{p_z} + \tfrac{1}{2}(N_{p_x} + N_{p_y})] \qquad (2.14^*)$$

where: $K_p = \tfrac{4}{5} <r^{-3}>_p$

N = orbital population.

It is evident from eq. (2.14) that if the three p orbitals are equally populated, then $q_{valence} = 0$. If $N_{p_z} > \tfrac{1}{2}(N_{p_x} + N_{p_y})$, then $q_{valence}$ is negative; if $N_{p_z} < \tfrac{1}{2}(N_{p_x} + N_{p_y})$, $q_{valence}$ will then be positive. As in the point charge treatment, a concentration of negative charge along the Z EFG axis gives

---

* This is the result of the Townes–Dailey treatment for quadrupole splittings. For a full discussion of the assumptions and approximations see reference 2.3.

a negative q. Similarly for d electrons:

$$q_{\text{valence}} = K_d[-N_{d_{z^2}} + N_{d_{x^2-y^2}} + N_{d_{xy}} - \tfrac{1}{2}(N_{d_{xz}} + N_{d_{yz}})] \quad (2.15)$$

where:

$$K_d = +\tfrac{4}{7}\langle r^{-3}\rangle_d$$

If the component orbitals of the t_{2g} and/or the e_g levels have equal populations, then $q_{\text{valence}} = 0$.

Separation of q_{valence}. It is convenient now to divide q_{valence} into two contributions:

$$q_{\text{valence}} = q_{\text{C.F.}} + q_{\text{M.O.}} \quad (2.16)$$

where: (a) $q_{\text{C.F.}}$ is the valence contribution considering a crystal field model with no overlap of ligand and metal orbitals,

and, (b) $q_{\text{M.O.}}$ is the valence contribution considering bonding between metal and ligands.

The $q_{\text{C.F.}}$ term dominates the q_{valence} contribution for transition metal ions such as Fe^{2+} high spin or Fe^{III} low spin in which the t_{2g} and/or e_g levels are not fully or half populated. The $q_{\text{M.O.}}$ term dominates the q_{valence} contribution for ions having symmetric (cubic or higher) electronic ground states (e.g. Sn^{IV} and Fe^{II} low spin). In other cases of interest, both terms may be of similar magnitude (e.g. Fe^0) and the separation given in eq. (2.16) is no longer useful.

To illustrate the value of the above separation, consider Fe^{II} low spin where the major part of the q_{valence} contribution can be explained considering $q_{\text{M.O.}}$; and Fe^{2+} high spin, where the major part of q_{valence} is usually due to $q_{\text{C.F.}}$. For Fe^{II} low spin $(t_{2g})^6$, (Fig. 2.6a), if there is no covalent bonding, $N_{d_{xy}} = \tfrac{1}{2}(N_{d_{xz}} + N_{d_{yz}})$, and $q_{\text{M.O.}}$, $q_{\text{C.F.}}$, and $q_{\text{valence}} = 0$. Any quadrupole splitting will arise from an unequal population of the d orbitals due to covalent bonding ($q_{\text{M.O.}}$), or a q_{lattice} contribution. Considering just the $q_{\text{M.O.}}$ contribution, consider a hypothetical species trans-$[Fe^{II}A_2B_4]^{++}$ (Figure 2.5a), where A and B are neutral ligands, and suppose that any Q.S. is due to the differences in π back bonding functions of A and B. Assume that A and B both have two perpendicular π orbitals which can accept electrons from the t_{2g} orbitals. Thus each t_{2g} d orbital will interact with four ligands (Fig. 2.7); each B interacts with d_{xy} and d_{xz} or d_{yz}, and each A with d_{yz} and d_{xz}. If A is a poorer π acceptor than B, then less electron density is withdrawn along the Z axes than along X or Y and $N_{d_{xy}} < \tfrac{1}{2}(N_{d_{xz}} + N_{d_{yz}})$, thus giving a negative contribution to $q_{\text{M.O.}}$; if A is a better π acceptor than B, then $N_{d_{xy}} > \tfrac{1}{2}(N_{d_{xz}} + N_{d_{yz}})$ and $q_{\text{M.O.}}$ is positive. The magnitude of the Q.S. will depend

Mössbauer spectroscopy

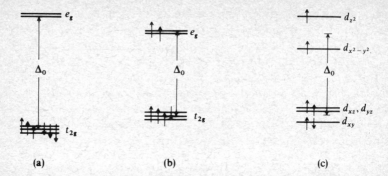

Fig. 2.6 d orbital energy level diagrams, and the ground state occupancy schemes for (a) Fe^{II} low spin in octahedral coordination, (b) Fe^{2+} high spin in octahedral coordination, and (c) Fe^{2+} high spin in a tetragonally distorted octahedral field.

on the difference in π accepting ability of A and B. Of course, σ donation is of great importance in determining quadrupole splitting values, and a detailed treatment which includes both σ and π bonding will be given in chapter 6.

The quadrupole splittings in many I, Te, Ru, Xe, and Sn compounds can be rationalized on a similar basis, and a detailed treatment of several cases will be discussed in chapter 6. Quadrupole splittings which arise from a $q_{M.O.}$ contribution normally vary little with temperature. Any small variation with temperature can be attributed to small changes in bond lengths.

In contrast to the above situation, the fourth electron in the t_{2g} level in Fe^{2+} high spin $(t_{2g}^4 e_g^2)$ (Fig. 2.6 b) normally gives rise to a large $q_{C.F.}$ term

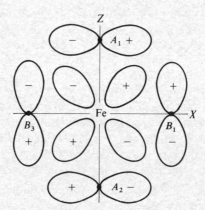

Fig. 2.7 The interaction of A and B π orbitals with Fe d_{xz} orbital in $trans$-FeA_2B_4. The d_{xy} orbital similarly interacts with the four B π orbitals in the xy plane.

which is very temperature dependent. If the Fe^{2+} is surrounded by a perfect octahedron of point charges, then the degeneracy of the t_{2g} levels is not removed, the extra electron equally populates the three t_{2g} orbitals and $q_{valence} = 0$. In this case, $N_{d_{xy}} = \frac{1}{2}(N_{d_{xz}} + N_{d_{yz}}) = 1\cdot33$. However, unlike Fe^{II} low spin compounds, Fe^{2+} compounds are inherently subject to a Jahn–Teller distortion which removes the degeneracy of the t_{2g} and e_g levels. If the axial ligands are compressed slightly relative to the equatorial ligands, then the fourth electron preferentially occupies the d_{xy} orbital, and a large positive $q_{C.F.}$ results (Figure 2.6 c and next section). If the separation $\Delta_3 \gg kT$, then $N_{d_{xy}} - \frac{1}{2}(N_{d_{xz}} + N_{d_{yz}}) = 1$ giving the large $q_{C.F.}$ contribution. However, usually $\Delta_3 \sim kT$ and the effective population $N_{d_{xy}} < 2$ due to the d_{xz} and d_{yz} orbitals being Boltzmann populated. For the above axial compression, $q_{C.F.}$ can be expressed as: (ref. 6.5)

$$q_{C.F.} = \tfrac{4}{7}\langle r^{-3}\rangle \left(\frac{1 - e^{-\Delta_3/kT}}{1 + 2e^{-\Delta_3/kT}}\right) \tag{2.17}$$

where: Δ_3 is the energy separation between d_{xy} and d_{xz}, d_{yz} (Fig. 2.6c)
k = Boltzmann's constant.
Thus the maximum value of $q_{C.F.} = \tfrac{4}{7}\langle r^{-3}\rangle$ is decreased due to thermal population; as T decreases, $q_{C.F.}$ increases ($e^{-\Delta_3/kT}$ decreases).

Spin orbit coupling, neglected in the above treatment, decreases the quadrupole splitting from the above value. The $q_{M.O.}$ and $q_{lattice}$ terms are usually much smaller than $q_{C.F.}$, but both usually decrease the observed quadrupole splitting from that expected just from the $q_{C.F.}$ term. This decrease can be illustrated as follows. Consider the axial compression discussed above which gives a positive $q_{C.F.}$. If the axial ligands are considered to be point charges, $q_{lattice}$ will be negative, as there will be a concentration of negative ligand charge along the Z EFG axis. Considering $q_{M.O.}$, if the axial ligand is a σ or π donor, then the negative ligand charge will again be concentrated along the Z axis and $q_{M.O.}$ is negative. Only if the axial ligands were π acceptors would a positive $q_{M.O.}$ contribution arise.

Thus, if Fe^{2+} high spin is surrounded by a perfect octahedron, no quadrupole splitting is observed. If a small distortion occurs such that d_{xy} is about 500 cm^{-1} below d_{xz}, d_{yz}, a large, temperature dependent quadrupole splitting will be observed. As the distortion increases, Δ_3 increases, the quadrupole splitting decreases and becomes less temperature dependent. For very large distortions from octahedral symmetry (such as square planar symmetry), it is possible to observe very small, temperature independent quadrupole splittings. In this case $q_{C.F.} \simeq -(q_{lattice} + q_{M.O.})$.

The above approach is applicable to transition metal ions such as Fe^{2+} high spin and Fe^{III} low spin which do not have filled or half filled valence

Mössbauer spectroscopy

shells or subshells, and thus give a $q_{C.F.}$ term. Uses of this approach will be discussed in chapter 6.

There are a number of ions for which the $q_{M.O.}$, $q_{C.F.}$ separation cannot be realistically justified. For example, consider Fe° compounds such as Fe(CO)$_5$. Fe° has the formal configuration d^8, and since the d_{z^2} and $d_{x^2-y^2}$ orbitals normally split in energy, a large $q_{C.F.}$ term arises from the two electrons in one or the other orbital. In addition to this, there will be a very large contribution from $q_{M.O.}$ from covalent bonding. It is usually very difficult to evaluate the magnitude of each contribution if indeed, such a separation is meaningful in such cases.

Units. Two simple calculations illustrate the units involved in calculating field gradients and quadrupole splittings from $q_{lattice}$ and $q_{valence}$ terms. Consider the expected ^{57}Fe quadrupole splitting from two point charges of magnitude e, 2·0 Å from the Fe atom (Figure 2.8). Since $3\cos^2\theta - 1$ is 2 for both ligands, $V_{zz} = -4(4·80 \times 10^{-10})/8·0 \times 10^{-24} = -2·4 \times 10^{14}$ e.s.u. cm^{-3}. Taking $Q = 0·2$ barns $= 0·2 \times 10^{-24}$ cm^2, then $\frac{1}{2}e^2 q_{lattice}Q (= \frac{1}{2}e\, V_{zz}Q) = -1·2 \times 10^{-20}$ ergs. Since 1 mm s^{-1} for ^{57}Fe corresponds to 7·69 × 10^{-20} ergs, (see Appendices), then $\frac{1}{2}e^2 q_{lattice}Q = -0·16$ mm s^{-1}. This value has to be multiplied by $(1 - \gamma_\infty)$ which is ~11, to give the calculated value of $\frac{1}{2}e^2 qQ = \sim -1·8$ mm s^{-1}.

The above calculation in SI units is demonstrated as follows. We take $e = 1·602 \times 10^{-19}$ coulombs, $r = 2·0 \times 10^{-10}$ m, $Q = 0·2 \times 10^{-28}$ m^2 and ε_0, the permittivity in vacuum $= 8·854 \times 10^{-12}$ Kg^{-1} m^{-3} s^{4}Å^2. $V_{zz} = -4(1·60 \times 10^{-19})/8·0 \times 10^{-30} = 8·0 \times 10^{10}$ coulombs m^{-3}. Then $\frac{1}{2}eV_{zz}Q/4\pi\varepsilon_0 = -1·2 \times 10^{-27}$ joules.

Using the expressions in Table 2.4, the expected quadrupole splitting from a single $3d$ or $4p$ electron in Fe^{2+} can be calculated as follows. Recent values of $\langle r^{-3} \rangle_{3d}$ and $\langle r^{-3} \rangle_{4p}$ are 4·73 a$_0^{-3}$ (or 4·73 a.u.) and 1·71 a$_0^{-3}$ respectively.

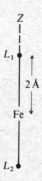

Fig. 2.8 Two negative point charges 2 Å from an Fe atom, both on the Z EFG axis.

Recalling that $1 a_0 = 0.529$ Å, $\langle r^{-3}\rangle_{3d}$ and $\langle r^{-3}\rangle_{4p}$ in cm^{-3} are 31.9×10^{24} cm^{-3} (31.9×10^{30} m^{-3}) and 11.6×10^{24} cm^{-3} (11.6×10^{30} m^{-3}) respectively. From these values we calculate $q_{3d} = -18.2 \times 10^{24}$ cm^{-3} (-18.2×10^{30} m^{-3}) for a $3d_{z^2}$ electron, and $q_{4p} = -9.3 \times 10^{24}$ cm^{-3} (-9.3×10^{30} m^{-3}) for a $4p_z$ electron. Thus the $4p$ contribution is about one half of the $3d$ contribution for equal orbital populations. The above q value for one $3d$ electron, leads to a $\frac{1}{2}e^2qQ$ value of -3.7 mm s^{-1}, taking $R = +0.32$ and $Q = 0.2$ barns. This value is close to the largest quadrupole splitting observed for Fe compounds, but it should be emphasized that there appears to be considerable uncertainty in this value due to uncertainties in $\langle r^{-3}\rangle$, R and Q.

2.3 Measurement of the sign of the quadrupole splitting

It is often important to know whether the sign of the quadrupole splitting is positive or negative, but as we have seen, for ^{57}Fe and ^{119}Sn it is not possible by taking the usual powder spectrum to obtain the sign of the quadrupole splitting. It is important to realize also that just as $\delta R/R$ can be positive or negative, so can Q. For ^{57}Fe, Q is positive, whereas for ^{119}Sn, Q is negative. Thus a measured positive quadrupole splitting would correspond to a positive q for ^{57}Fe but a negative q for ^{119}Sn.

There are two methods which are normally used to obtain the sign of q. The first involves obtaining spectra of a sample with all crystallites oriented in one known way relative to the direction of the gamma beam; the second involves measuring the spectrum of a polycrystalline sample at 4 K in a large magnetic field.

For a random polycrystalline sample, the intensity of the two peaks should be very nearly the same. For a single crystal, the intensity ratio of the two lines is no longer 1:1, but varies with the orientation of the crystal relative to the direction of the gamma beam. The angular dependence of the intensity of the two lines can be expressed as:

Transition	Angular Dependence
$\pm\frac{3}{2} \to \pm\frac{1}{2}$	$1 + \cos^2\theta$
$\pm\frac{1}{2} \to \pm\frac{1}{2}$	$\frac{5}{3} - \cos^2\theta$

where θ is the angle between the Z EFG axis and the direction of the gamma ray. It is necessary to know the orientation of the molecular axes with respect to the crystal axes to determine θ. If we have a single crystal or crystals for which it is known that $\theta = 0°$, then the intensity ratio $I_{3/2}:I_{1/2}$ approaches 3; if $\theta = 90°$, the intensity ratio approaches 3:5. For $\theta = 0°$, if the most intense line (the $\frac{3}{2}$ line) is at positive velocities, the sign of the quadrupole splitting (and q for positive Q) is positive. The limiting ratios of 3:1 and 3:5 are not approached very closely for two main reasons: first, the gamma beam is usually not well collimated, and θ varies over a small range of angles about $0°$

Mössbauer spectroscopy

or 90°; second, the f factor may be anisotropic (the Goldanskii–Karyagin effect) which also causes the powder area ratio to vary from 1:1.

On application of a large magnetic field to a powder sample, the degeneracy of the nuclear levels is completely removed (see section 2.5). The EFG axes take all orientations with respect to the magnetic field, and a large number of superimposed spectra are observed. For zero or small η, the two-line zero-field spectrum splits into a doublet and a triplet (Fig. 2.9), with the doublet

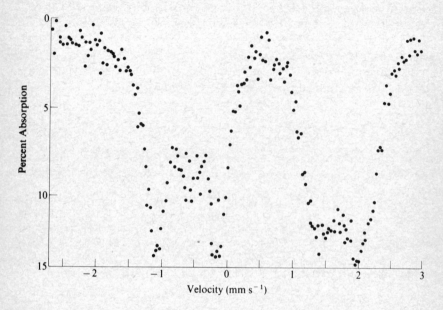

Fig. 2.9 Effect of a magnetic field on an ^{57}Fe powder spectrum—the Mössbauer spectrum of ferrocene at 4·2 K in an applied longitudinal magnetic field of 40 Kgauss. The doublet lies to +ve velocities and V_{zz} and q are positive (R. L. Collins, *J. Chem. Phys.*, **42**, 1072, 1965).

being due to the $+\frac{1}{2} \to +\frac{3}{2}$ and $-\frac{1}{2} \to -\frac{3}{2}$ transitions. For ^{57}Fe, then, if the doublet lies to positive velocity, the sign of the quadrupole splitting and q is positive. If η approaches one, the spectrum goes from a doublet–triplet structure to a symmetric triplet–triplet structure, indicating that the sign is indeterminate. Using detailed computation, an estimate of η can be made. Orientation of the crystallites or an anisotropic f factor can markedly alter the spectrum and lead to difficulties in detailed interpretation, especially for small quadrupole splittings (<0.50 mm s^{-1}), but the sign can still usually be obtained. A detailed discussion of this technique is given in reference 2.4.

For ^{119}Sn, a doublet–quartet structure is observed, and the sign of the quadrupole splitting is positive (q negative) if the doublet is at positive velocities (ref. 2.5).

2.4 Quadrupole hyperfine parameters for $I > \tfrac{3}{2}$

Many Mössbauer elements such as iodine, tungsten, and the rare earths give complex spectra arising from the many possible transitions from the large spins of ground and/or excited states. It seems important here to indicate how the quadrupole hyperfine parameters can be obtained from such spectra. For a proper treatment, standard quantum mechanics is required, but in this section, a simplified example will be considered which requires no quantum mechanical background. In reference 2.6, the complete analytical treatment for all common nuclear spins is considered. Using this method, all the quadrupole parameters can easily be obtained for spectra in which most of the lines can be resolved.

The Hamiltonian representing the interaction of a nucleus having spin I and quadrupole moment Q with an EFG may be written as:

$$\mathcal{H} = \frac{e^2qQ}{4I(2I-1)}[3I_z^2 - I(I+1) + (\eta/2)(I_+^2 + I_-^2)] \qquad (2.18)$$

where $I_\pm = I_x \pm iI_y$ and I_x, I_y and I_z are the nuclear spin component operators. For $\eta = 0$, the energy levels of a nucleus are given exactly by:

$$E_m = \frac{e^2qQ[3m_I^2 - I(I+1)]}{4I(2I-1)} \qquad (2.19)$$

where $m_I = I, I-1 \ldots -I$.

For $I = \tfrac{3}{2}$, $E_{\pm 3/2} - E_{\pm 1/2} = \tfrac{1}{2}e^2qQ$ as is given in eq. (2.7). For $\eta \neq 0$, the energy levels are generally *not* given by the equation widely quoted.

$$E_m \neq \frac{e^2q\,Q[3m_I^2 - I(I+1)]}{4I(2I-1)}(1 + \eta^2/3)^{1/2} \qquad (2.20)$$

For $\eta \neq 0$, this equation is only good for deriving the quadruple splitting for $I = \tfrac{3}{2}$ (equation (2.7)). A general case for $\eta \neq 0$ is considered in reference 2.6.

Consider ^{129}I. The ground and excited states have spins of $\tfrac{7}{2}$ and $\tfrac{5}{2}$ respectively and Q is negative for both. Since m_I^2 appears in eq. (2.19), the energy levels are doublets (Kramer's doublets) as seen in Figure 2.10 for $e^2qQ = -Ve$. The usual selection rule $\Delta m_I = 0, \pm 1$ holds, so that eight transitions are observed with intensities given by the squares of the Clebsch–Gordon coefficients (circled in Fig. 2.10). As in the $I = \tfrac{3}{2}$ case considered earlier in this chapter, the centre of gravity of the excited and ground nuclear energy levels is not displaced by the quadrupole interaction, i.e.

$$\sum_{i=1\to 3}{}^eE_i = {}^eE_a \quad \text{and} \quad \sum_{j=1\to 4}{}^gE_j = {}^gE_a$$

where the superscripts e and g refer to excited and ground states. By substituting the values of I and m_I into eq. (2.19), the energy levels given in

Mössbauer spectroscopy

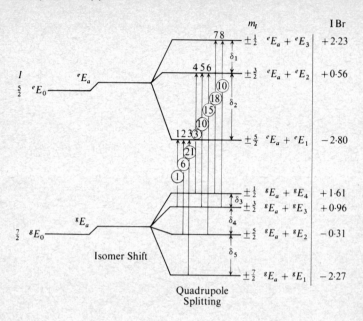

Fig. 2.10 Isomer shift and quadrupole splitting in ^{129}I. Numbers refer to lines in Fig. 2.11; bracketed numbers are the Clebsch–Gordon coefficients. Figures on the right give the ^eE and ^gE values for the IBr spectrum (Table 2.6 and Fig. 2.11) in cm s^{-1}.

Table 2.5 can be calculated in terms of U_I, where

$$U_I = \frac{e^2qQ}{4I(2I-1)} \quad (2.21)$$

Note that $\delta_1(^eE_3 - {}^eE_2) = \frac{1}{2}\delta_2(^eE_2 - {}^eE_1)$, and $\delta_3:\delta_4:\delta_5 = 1:2:3$. (Fig. 2.10). The eight lines observed can be expressed as:

$$L_K \atop K=1\to 8 = [(^eE_a + {}^eE_i) - (^gE_a + {}^gE_j)] - [^eE_s - {}^gE_s]$$

Rearranging:

$$L_K = (^eE_i - {}^gE_j) + [(^eE_a - {}^gE_a) - (^eE_s - {}^gE_s)]$$

where E_a and E_s are the absorber and source energy levels respectively. The square bracketed term is just the centre shift, and L_K can now be expressed as:

$$L_K = {}^eE_i - {}^gE_j + \text{C.S.} \quad (2.22)$$

Consider now the spectrum of IBr (Fig. 2.11). The approximate line positions are given in Table 2.6, and the lines are numbered as in Figure 2.10. The lines could be assigned using the Clebsch–Gordon coefficients. For example, line 1 is the least intense, while line 3 is the most intense. The

Table 2.5 Energy levels for the ground and excited levels of ^{129}I in terms of U^* ($\eta = 0$)

Excited state		Ground state	
Level	Energy	Level	Energy
$^eE_3$	$-8U_{5/2}$	$^gE_4$	$-15U_{7/2}$
$^eE_2$	$-2U_{5/2}$	$^gE_3$	$-9U_{7/2}$
$^eE_1$	$+10U_{5/2}$	$^gE_2$	$+3U_{7/2}$
		$^gE_1$	$+21U_{7/2}$

* $U = e^2qQ/4I(2I - 1)$.

lines can also be assigned by subtracting one line position from another. A set of constant differences may be set up. For example, $L_5 - L_1 = L_6 - L_2 = 3 \cdot 36$; and $L_7 - L_4 = L_8 - L_5 = 1 \cdot 67$. Similarly $L_2 - L_1 = L_6 - L_5 = 1 \cdot 27$; $L_5 - L_4 = L_8 - L_7 = 0 \cdot 65$ and $L_3 - L_2 \simeq L_6 - L_4 \simeq 1 \cdot 95$ (all values in cm s^{-1}). From the relationship of the values stated previously, it is apparent that $\delta_1 = 1 \cdot 67$, $\delta_2 = 3 \cdot 36$, $\delta_3 = 0 \cdot 65$, $\delta_4 = 1 \cdot 27$ and $\delta_5 = 1 \cdot 95$. Then using the following equations for $I = \frac{5}{2}$ relative to ^eE_a:

$$^eE_1 + {}^eE_2 + {}^eE_3 = 0 \quad (2.23a)$$

$$^eE_3 - {}^eE_2 = 1 \cdot 67 \quad (2.23b)$$

$$^eE_2 - {}^eE_1 = 3 \cdot 36 \quad (2.23c)$$

Solving eqs. (2.23), the energy level diagram shown in Fig. 2.10 for IBr can be constructed for the $I = \frac{5}{2}$ state, and using an analogous set of equations, the $I = \frac{7}{2}$ energy levels can be constructed.

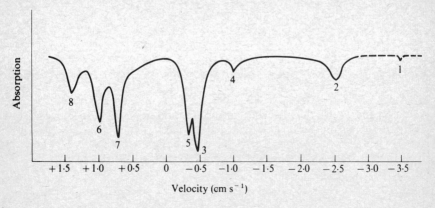

Fig. 2.11 Mössbauer spectrum of IBr (M. Pasternak and T. Sonnino, *J. Chem. Phys.*, **48**, 1997, (1968)).

Mössbauer spectroscopy

Table 2.6 Approximate line positions for the Mössbauer spectrum of IBr

Line	Velocity (cm s^{-1})	Line	Velocity (cm s^{-1})
1	−3·63	5	−0·27
2	−2·36	6	+1·00
3	−0·40	7	+0·75
4	−0·92	8	+1·40

(M. Pasternak and T. Sonnino, *J. Chem. Phys.*, **48**, 1997 (1968).)

From eq. (2.22), the centre shift value can be calculated. For line 3, $L_K = -0.40$, $^eE_1 = -2.80$ and $^gE_1 = -2.27$. Substituting, the centre shift equals 0.13 cm s^{-1}. The value of e^2qQ can now be calculated for both ground and excited states using Table 2.5 and eq. (2.21). From Table 2.5,

$$^eE_3 - {}^eE_2 = -6U_{5/2} = 1.67$$

and $U_{5/2} = -0.28$. Similarly for the ground state

$$^gE_4 - {}^gE_3 = -6U_{7/2} = 0.65$$

and $U_{7/2} = -0.11$. From eq. (2.21), $e^2q_{5/2}Q_{5/2} = 40U_{5/2}$, and $e^2q_{7/2}Q_{7/2} = 84U_{7/2}$. This then gives

$$e^2q_{5/2}Q_{5/2} = -11.2; \quad e^2q_{7/2}Q_{7/2} = -9.2 \quad \text{(in cm s}^{-1}\text{)}$$

and since $q_{5/2} = q_{7/2}$, $Q_{5/2}/Q_{7/2} \simeq 1.2$.

Thus e^2qQ, the centre shift and the ratio of the quadrupole moments can be readily determined for $\eta = 0$. If $\eta \neq 0$, $\delta_1 \neq \frac{1}{2}\delta_2$, and the full treatment given in reference 2.6 will have to be used.

For IBr, the signs of e^2qQ are negative. A positive e^2qQ is easily distinguished from a negative one in at least two ways. First, the small, well spaced lines 1, 2, and 4 are at positive velocities for $e^2qQ = +Ve$, but at negative velocities for $e^2qQ = -Ve$. Second, after assigning the differences by subtracting line positions, if the incorrect sign is chosen, the line positions calculated from eq. (2.22) will not agree with the spectral line positions.

2.5 Magnetic splitting

In later chapters of this book, a few magnetic spectra of minerals will be given, but the detailed interpretation of magnetic splittings will not be considered. It is appropriate here to give a very brief account of magnetic hyperfine interactions in the case of ^{57}Fe.

As discussed in chapter 1, a nucleus having a spin I has a magnetic dipole moment U_N, given by the equation:

$$U_N = g_N \beta_N I \tag{1.11}$$

Mössbauer parameters and theory

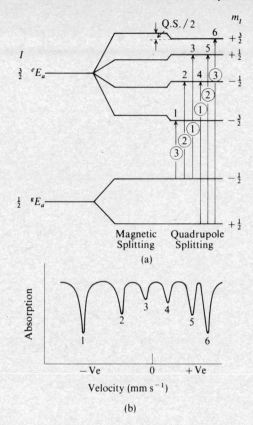

Fig. 2.12 Magnetic and quadrupole splitting in ^{57}Fe. (a) Energy level diagram for combined magnetic and quadrupole splitting, (b) the resultant Mössbauer spectrum. The circled numbers in (a) are the expected intensities for a random sample, and the other numbers refer to the lines in (b).

This magnetic dipole moment interacts with local or applied magnetic fields at the nucleus, and the degeneracy of the nuclear energy levels is completely removed (Fig. 2.12). For ^{57}Fe, the g_N values of ground and excited states have different signs, and the selection rules $\Delta m_I = 0, \pm 1$ give rise to a symmetric six-line spectrum. The centre shift is given by the centre of gravity of the six peaks.

The line widths of the six peaks are in general equal, but the intensities are very different. The intensities are given by

$$I_1 = I_6 = 3(1 + \cos^2 \theta)$$
$$I_2 = I_5 = 4 \sin^2 \theta$$
$$I_3 = I_4 = 1 + \cos^2 \theta$$

where θ is the angle between the effective H and the direction of propogation of the radiation. For a sample (such as in Fig. 2.12) in which the magnetic domains are randomly oriented, the area ratio of the six lines is $3:2:1:1:2:3$.

In most magnetic substances, the iron atom does not occupy a site with cubic point symmetry, and a quadrupole splitting as well as the magnetic splitting will arise. In this case there are five possible unknowns: e^2qQ and η; and a magnetic field H at an orientation to the principle axes of the EFG defined by the angles α, β and γ. The latter three are related by $\cos^2\alpha + \cos^2\beta + \cos^2\gamma = 1$ so that only two are independent. In the simplest case, when H is parallel to the Z EFG axes and $\eta = 0$, Q.S. $= \frac{1}{2}[(L_6 - L_5) - (L_2 - L_1)]$ (Fig. 2.12). For the general case, the problem becomes much more complex. Approaches to the problem are given in reference 2.6. The analytical approach appears to be the easiest to use and gives the most information for many cases of interest.

2.6 Line shape, width, and area

Resonant absorption shows a characteristic energy dependence of the form:

$$I(E) = \frac{\Gamma_{ex}}{2\pi}\left[\frac{1}{(E - E_t)^2 + (\frac{1}{2}\Gamma_{ex})^2}\right] \qquad (2.24)$$

where Γ_{ex} = the experimentally determined full width at half maximum and E_t = the nuclear transition energy. For ^{57}Fe, this equals 14.4 keV. This distribution is said to show a Breit–Wigner or Lorentzian shape. Fortunately, the Mössbauer line shapes are remarkably Lorentzian in character and make it possible to account for any small shoulders or lumps in a complex spectrum using detailed computer processing (chapter 3.3).

The natural line width of a Mössbauer line is determined by the half life of the excited nuclear state and the Heisenberg uncertainty principle as seen in chapter 1. The line is broadened by several effects, one of which is so called thickness broadening. The experimental full width at half height can be expressed as:

$$\Gamma_{ex} = \Gamma_a + \Gamma_s + 0.27\,\Gamma_H X \qquad (2.25)$$

where: $\Gamma_a + \Gamma_s$ is the width of Γ_{ex} extrapolated to $X = 0$.

Γ_H is the natural line width determined by the lifetime of the excited nuclear state and $X = nf_a\sigma_o$ where: n is the number of atoms of the Mössbauer isotope per cm^2, f_a is the recoil free fraction of the absorber and σ_o is the maximum cross section at resonance. For iron this is 2.35×10^{-18} cm^2. By measuring the widths of the absorber lines for different absorber thicknesses, f_a may be determined using equation 2.25.

Mössbauer parameters and theory

The area A under a peak for a 'thin' Lorentzian one line absorber and a single-line source of arbitrary width may be expressed as:

$$A = \tfrac{1}{2}\pi f_a f_s \sigma_o \Gamma_{ex} G(X) n \tag{2.26}$$

where f_s is the recoil free fraction of the source and $G(X)$ approaches one as X approaches zero, i.e. as $n f_a \sigma_o \to 0$. As X increases, $G(X)$ decreases. Several other expressions for the area have been derived, but the above one is convenient because it contains n explicitly.

Theoretically, it should be possible to determine n from A, but in practice there are many difficulties. The area A has to be corrected for non-Mössbauer radiation which enters the final stage of the counting equipment. Such radiation normally includes X-rays, Compton scattered gamma rays, and gamma rays which are reemitted after being absorbed. It is also difficult to determine accurate values of f_a, f_s and $G(X)$, and care must be taken to prepare homogeneous absorbers. Many of these problems are minimized or eliminated when the ratio of the number of iron atoms in two different sites are determined from the ratio of areas i.e. $A_2/A_1 \propto n_2/n_1$. The use of area ratios will be discussed in detail in chapters 8 and 9.

As the thickness X of the sample increases, the line shapes begin to deviate from Lorentzian shape and $G(X)$ decreases markedly from unity—a process known as saturation. For very large thicknesses $A \propto n^2$. For ^{57}Fe, as long as a sample contains less than 10 mg cm^{-2} of natural iron, any deviation from Lorentzian shape cannot be detected after very extensive curve fitting.

To illustrate the effect of saturation, consider a sample containing two different types of iron atoms Fe$_A$ and Fe$_B$ with three times as much Fe$_A$ as Fe$_B$. Since $X_A > X_B$, then $G(X_A) < G(X_B)$. Using eq. (2.26) and assuming that $\Gamma_A = \Gamma_B$ and $f_A = f_B$, then $A_A/A_B < 3$, and this area ratio will decrease as the thickness of the sample increases. This effect is illustrated in Table 8.5.

2.7 Characteristics of a useful Mössbauer isotope

The great majority of Mössbauer work has been done on ^{57}Fe and ^{119}Sn, yet there are at least thirty-three elements which show the effect, and it has been predicted for at least seventeen others. Table 2.7 lists important properties of most of the potentially useful isotopes, apart from the lanthanides and actinides (see reference 2.7). Referring to Table 2.7, we will consider criteria which make a particular isotope useful for Mössbauer research.

First and foremost, it is necessary to have a radioactive isotope which emits a gamma ray of less than 150 keV. As discussed in chapter 1 (eq. 1.24), the lower the gamma ray energy, the larger the Mössbauer f factor. It is also generally true that as E_γ increases, the spectrum must be recorded with source and/or absorber cooled—often to 4 K. ^{57}Fe and ^{119}Sn, with low gamma energies, often give reasonable spectra at room temperature;

Mössbauer spectroscopy

Table 2.7 Mössbauer isotopes of chemical and mineralogical interest‡

Isotope	Gamma energy (keV)	Half life* of precursor	Half width (mm s^{-1})	Spin	Q† (Barns)	σ_0 (10^{-19} cm^2)	Natural abundance (%)	$10^4 \frac{\delta R}{R}$
^{57}Fe	14.4	270 d	0.19	$\frac{1}{2} \to \frac{3}{2}$	+0.3	23.5	2.19	−18 ± 4
^{61}Ni	67.4	1.7 h; 3.3 h	0.78	$\frac{3}{2} \to \frac{5}{2}$	0.13	7.21	1.19	
^{67}Zn	93.3	60 h; 78 h	3.13 × 10^{-4}	$\frac{5}{2} \to \frac{3}{2}$	+0.17	1.18	4.11	
^{73}Ge	67.0	76 d	2.19	$\frac{9}{2} \to \frac{7}{2}$	−0.26	3.54	7.76	
^{83}Kr	9.3	83 d, 2.4 h	0.20	$\frac{9}{2} \to \frac{7}{2}$	+0.44	18.9	11.55	+4 ± 2
^{99}Ru	90	16.1 d	0.15	$\frac{5}{2} \to \frac{3}{2}$	> 0.15	1.42	12.72	
^{107}Ag	93.1	6.6 h	6.68 × 10^{-11}	$\frac{1}{2} \to \frac{7}{2}$	—	0.54	51.35	
^{119}Sn	23.9	245 d	0.62	$\frac{1}{2} \to \frac{3}{2}$	−0.07	13.8	8.58	+3.3 ± 1
^{121}Sb	37.2	76 y	2.10	$\frac{5}{2} \to \frac{7}{2}$	−0.29	2.04	57.25	−8.5 ± 3
^{125}Te	35.5	58 d	4.94	$\frac{1}{2} \to \frac{3}{2}$	±0.19	2.72	6.99	+1
^{129}I	27.8	33 d; 70 m	0.63	$\frac{7}{2} \to \frac{5}{2}$	−0.55	3.97	0	+3
^{129}Xe	39.6	1.6 × 10^7 y	6.84	$\frac{1}{2} \to \frac{3}{2}$	−0.41	1.95	26.44	0.3
^{133}Cs	81.0	7.2 y	0.54	$\frac{7}{2} \to \frac{5}{2}$	−0.003	1.02	100	
^{177}Hf	113.0	6.7 d, 56 h	4.66	$\frac{7}{2} \to \frac{9}{2}$	+3	1.20	18.50	
^{181}Ta	6.25	140 d, 45 d	6.48 × 10^{-3}	$\frac{7}{2} \to \frac{9}{2}$	+4.2	17.2	99.99	
^{182}W	100.1	115 d	2.00	$0 \to 2$	−1.87	2.46	26.41	+1.3
^{187}Re	134.2	23.8 h	203.8	$\frac{5}{2} \to \frac{7}{2}$	+2.6	0.54	62.93	
^{186}Os	137.2	90 h	2.37	$0 \to 2$	+1.54	2.89	1.64	
^{193}Ir	73.1	32 h	0.59	$\frac{3}{2} \to \frac{1}{2}$	+1.5	0.30	62.7	+0.6
^{195}Pt	98.7	183 d	17.30	$\frac{1}{2} \to \frac{3}{2}$	—	0.63	33.8	
^{197}Au	77.3	65 h, 20 h	1.85	$\frac{3}{2} \to \frac{1}{2}$	+0.58	0.44	100	±3

* d = days, h = hours, y = years, m = minutes.
† The ground state quadrupole moment where both ground and excited states have $I > \frac{1}{2}$.
‡ Much of these data comes from: *Mössbauer Effect Data Index*, ed. J. G. Stevens and V. E. Stevens, Plenum Press (1970).

whereas for ^{182}W and ^{195}Pt, liquid helium temperatures (4 K) are usually required to observe absorptions.

The Mössbauer gamma ray should be well separated in energy from other photons, so that a great deal of noise (non-Mössbauer radiations) is not fed into the final stage of counting. In some cases special counters have to be used (chapter 3). For example, for ^{193}Ir, the 73 keV gamma ray cannot be resolved from other radiations using the common scintillation or proportional counters. A good Li drifted germanium counter can resolve the 73 keV gamma ray from other photons, making it possible to feed mostly Mössbauer gamma rays to the multichannel analyzer (section 3.1).

For a large effect, the natural abundance of the Mössbauer isotope should be large because, as seen in eq. (2.26), A is proportional to n. Enrichment of compounds is of course possible, but expensive and time-consuming. For example, ^{129}I is radioactive, and all compounds must be enriched with this rather expensive isotope. The ^{129}I must then be recovered after nearly every preparation.

For a large effect, consideration must be given to the cross-section σ_o for absorption of the Mössbauer gamma ray. This is given by:

$$\sigma_o = \frac{\lambda^2}{2\pi} \frac{(1 + 2I_e)}{(1 + 2I_g)} \frac{1}{1 + \alpha} \qquad (2.27a)$$

where: I_e and I_g are the nuclear spins of the excited and ground states respectively

α is the internal conversion coefficient (chapter 1).

Expressing λ in terms of E_γ, we obtain:

$$\sigma_o(\text{cm}^2) = \frac{2 \cdot 446 \times 10^{-15}}{E_\gamma^2(\text{keV})} \frac{(1 + 2I_e)}{(1 + 2I_g)} \frac{1}{1 + \alpha} \qquad (2.27b)$$

For a large σ_o, and a large Mössbauer absorption, it is apparent from eq. (2.27b) that E_γ and α should be small. It is apparent from Table 2.7 that σ_o is very large for ^{57}Fe and ^{119}Sn compared to other isotopes having large E_γ such as ^{195}Pt and ^{197}Au.

For convenience and economy, it is desirable that the half life of the precursor is long. ^{57}Co and ^{119m}Sn, the precursors for iron and tin Mössbauer, have convenient half lives; while ^{61}Co and ^{61}Cu, the precursors for nickel Mössbauer, have half lives of 1·7 hours and 3·3 hours respectively. These latter half lives are too short for routine experiments, unless irradiation facilities are available nearby to regenerate the source cheaply.

Given that we have a Mössbauer isotope which gives a large effect, and a source of reasonable half life, the line positions must be sensitive to the chemical environment and any changes in line position must be measurable. The sensitivity depends on the magnitude of the half width compared to the

Mössbauer spectroscopy

magnitude of the centre shift and quadrupole splitting. For iron compounds, the centre shift and quadrupole splitting are much larger than the line widths. ^{119}Sn spectra are somewhat less well resolved because of the much larger half width (0·62 mm s^{-1}). Isotopes such as ^{187}Re and ^{195}Pt will have very little chemical interest since their half widths (203·8 mm s^{-1}, and 17·30 mm s^{-1}) are so large that any shifts or splittings will not be seen. One broad unshifted line will always be observed. For an isotope to be useful, the half width should be less than ∼8 mm s^{-1}. On the other hand, isotopes such as ^{67}Zn and ^{181}Ta have line widths which are too small for convenience. In order to observe peaks for these isotopes, velocities accurate to about 10^{-5} mm s^{-1} must be obtained. With vibration problems, this is extremely difficult. The narrowest useful line width appears to be about 0·05 mm s^{-1}.

Finally, it is usually convenient to have a fairly simple spin system, although as discussed earlier in this chapter, more information can be obtained for ^{129}I than ^{57}Fe or ^{119}Sn. However, the quadrupole splittings are very large in ^{129}I, and all the eight lines can be resolved. With other isotopes such as ^{121}Sb ($I_e = \frac{7}{2}$; $I_g = \frac{5}{2}$), the lines strongly overlap making the analysis and extraction of the quadrupole parameters more difficult.

Apart from ^{67}Zn, ^{107}Ag, ^{181}Ta, ^{187}Re and ^{195}Pt, most of the isotopes in Table 2.7 should be useful for chemical or mineralogical studies, although some of the problems outlined above make some of the other isotopes difficult and rather expensive to use conveniently.

Problems

1. For the species FeF$_4^=$, calculate the expected quadrupole splitting from a $q_{lattice}$ contribution for:
   (a) a tetrahedral arrangement of F$^-$ about Fe^{2+}
   (b) a square planar arrangement of F$^-$ about Fe^{2+}
   Take $Q_{57Fe} = 0.3$ barns and $r_{Fe-F} = 2$ Å. How would $q_{valence}$ affect the above quadrupole splittings?

2. Johnson, Marshall and Perlow [*Phys. Rev.* **126**, 1503, (1962)] were among the first workers to estimate Q_{57Fe} using the measured $\frac{1}{2}e^2qQ$ value of -3.70 mm s^{-1} for Fe^{2+}SiF$_6$.6H$_2$O. Taking the ground state orbital to be d_{z^2} and neglecting spin orbit coupling and Boltzmann population of other orbitals, calculate Q_{57Fe} taking $\langle r^{-3} \rangle_{3d} = 5.1\, a_0^{-3}$. Neglect any $q_{lattice}$ contributions.

3. A radioactive atom of mass 40 emits a 100 keV gamma ray. The excited nuclear state of the above isotope has a $t_{1/2}$ of 10^{-8} sec, and $I_{excited} = \frac{7}{2}$ and $I_{ground} = \frac{5}{2}$. Take $Q_{excited}/Q_{ground} = +6$. Typical quadrupole splittings (e^2qQ_{ex} values) are -3 mm s^{-1}. What problems (if any) would you

envisage in extracting quadrupole splittings from compounds of the above element?

4. The ^{59}Co N.Q.R. frequencies in $Cl_3SnCo(CO)_4$ (^{59}Co has $I = \frac{7}{2}$) are 35·02 MHz ($\pm\frac{5}{2} \rightarrow \pm\frac{7}{2}$), 23·37 MHz ($\pm\frac{3}{2} \rightarrow \pm\frac{5}{2}$) and 11·68 MHz ($\pm\frac{1}{2} \rightarrow \pm\frac{3}{2}$). [D. D. Spencer, J. L. Kirsch and T. L. Brown, *Inorg. Chem.* **9**, 235, (1970)]. Taking the experimental error in the frequencies as ± 0.05 MHz, is $\eta = 0$? Calculate e^2qQ.

5. The mineral gillespite $BaFeSi_4O_{10}$ contains Fe^{2+} high spin coordinated to four oxygens in a square planar arrangement. The Z EFG axis lies along the four fold molecular axis. Gillespite forms platelets which have this four fold axis perpendicular to the face of the platelet.

   The gillespite sample is mounted in the Mössbauer spectrometer such that the γ beam is parallel to the Z EFG axis ($\theta = 0$). The two line ^{57}Fe spectrum gives the high velocity line being substantially more intense than the low velocity line. What is the sign of e^2qQ? The electronic spectra of gillespite strongly indicate that the Fe ground state d orbital is d_{z^2}. Account for the sign of e^2qQ considering both $q_{valence}$ and $q_{lattice}$.

6. For the two isomers of MA_3B_3, write the point charge expressions for the three diagonal components of the EFG about M. What will the sign of the quadrupole splitting be in the two isomers?

7. Me_2SnCl_2 is known to have a polymeric structure with octahedral coordination about the Sn atom.

```
        Me           Me
   Cl   |     Cl     |
    \   |   / \      |  /
       Sn       Sn
    /  |  \   / |  \
   Cl  |   Cl    |
       Me         Me
```

Taking the Z EFG axis through the Me–Sn bond, and remembering that Me is a much better donor than Cl (or Cl is more electronegative than Me), what sign would you expect for q and e^2qQ?

8. $(NEt_4)_2Fe^{2+}Cl_4$ was found to have a large temperature dependent quadrupole splitting. Would this splitting be expected if the Cl atoms form a perfect tetrahedron about the Fe atom?

   Given the quadrupole splittings 2·49 mm s^{-1} at 80 K and 1·63 mm s^{-1} at 190 K, calculate the energy separation between the d_{z^2} and $d_{x^2-y^2}$ orbitals. Hint: See equations 6.26 and 6.28, chapter 6.

9. From the ^{129}I Mössbauer spectrum of ICN in *J. Chem. Phys.*, **48**, 2009, (1968), calculate $e^2qQ_{127_I}$ and the centre shift using the method of differences described in this chapter and the following data: line 2 at -2.20 cm s^{-1}, line 1 at -3.35 cm s^{-1}, and $\eta = 0$.

References

2.1 N. E. Erickson in *The Mössbauer Effect and its Applications in Chemistry*, American Chemical Society Publication, 1967.
2.2 J. K. Lees and P. A. Flinn, *J. Chem. Phys.*, **48**, 882 (1968).
2.3 E. A. C. Lucken, *Nuclear Quadrupole Coupling Constants*, Academic Press, 1969.
2.4 R. L. Collins and J. C. Travis, *Mössbauer Effect Methodology*, Plenum Press, ed. I. J. Gruverman, 1967, vol. 3.
2.5 T. C. Gibb, *J. Chem. Soc.* (A), 2503 (1970).
2.6 P. G. L. Williams and G. M. Bancroft, *Mössbauer Effect Methodology*, Plenum Press, ed. I. J. Gruverman, 1971, vol. 7.
2.7 *Mössbauer Effect Data Index*, ed. J. G. Stevens and V. E. Stevens, Plenum Press, 1970.

Bibliography

G. K. Wertheim, *Mössbauer Effect: Principles and Applications*, Academic Press, 1964.
T. C. Gibb and N. N. Greenwood, *Mössbauer Spectroscopy*, Chapman and Hall, 1971.
L. May ed., *An Introduction to Mössbauer Spectroscopy*, Plenum Press, 1971.

3. Experimental techniques

In comparison to many spectroscopic experiments, the basic Mössbauer equipment is rather simple and inexpensive. A block diagram of the equipment is shown in Fig. 3.1. It consists basically of a drive mechanism to

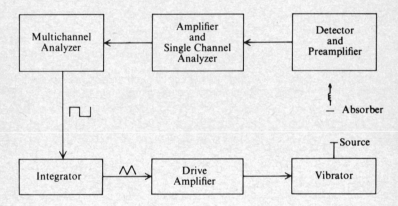

Fig. 3.1 Block diagram of a typical Mössbauer spectrometer.

impart a known and controllable velocity to a radioactive source such as ^{57}Co which emits low energy Mössbauer gamma rays, a gamma ray detector with associated amplifying and sorting equipment, and some sort of data storing device such as a multichannel analyzer. In this chapter, we will discuss characteristics of different types of the above equipment, and also discuss other important experimental problems such as absorber preparation, source to detector distances, calibration, and background corrections. A discussion of cryostats and heaters will be omitted. Excellent reviews of such equipment are given in references 3.1 and 3.2.

3.1 Basic Mössbauer equipment

The drive mechanism. As mentioned in the previous chapters, a Mössbauer spectrum consists of a plot of the number of gamma ray photons transmitted through an absorber as a function of the instantaneous relative velocity of

Mössbauer spectroscopy

the source with respect to an absorber. Many types of mechanisms have been used to impart velocities to either the source or the absorber, but these fall into two main categories: purely mechanical drives such as rotating discs or cams, and electromechanical devices such as loudspeaker coils or vibrators. Drives are normally of two types: constant velocity, in which a range of fixed velocities are imparted, one by one, to the source-absorber system, and counts are recorded for a given length of time at each velocity; and constant acceleration, in which a range of velocities is scanned linearly and repetitiously with the counts being recorded in a multichannel analyzer such that the velocity increment per channel is a constant.

The mechanical drives appear to have three advantages. They require very little costly electronics and can be built by most laboratory workshops, the best types can provide very high absolute accuracy, and it is easier to cool the source, since these drives are normally very sturdy. However, certain disadvantages usually outweigh the advantages. It is more difficult to eliminate extraneous vibrations than in the electromechanical drives, and mechanical wear may eventually cause difficulties. In addition, it is usually only possible to use these drives in constant velocity motion. This requires that the spectrum be built up point by point, and drifts in the counting equipment may then become important.

Electromechanical devices are most often used to impart the Döppler velocity to the source or absorber, and many such systems are now

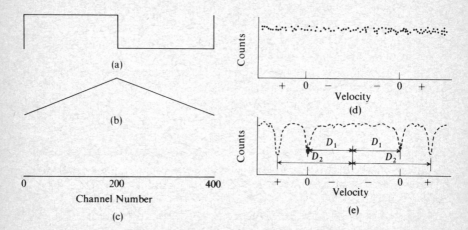

Fig. 3.2 Synchronization of multichannel analyzer and vibrator sweeps. (a) Square wave from the multichannel analyzer with leading edges at channels 0 and 200, (b) Symmetric saw tooth from integrated square wave ready to be fed into the vibrator, (c) Multichannel analyzer sweep in the time mode with no counts being accumulated, (d) Spectrum with no absorber. The scatter of points reflects the statistical error in a count, (e) Mössbauer spectrum. A mirror image spectrum is obtained, with $D_1 = D_1$ and $D_2 = D_2$.

commercially available. The vibrator is driven by a symmetric sawtooth or a rounded asymmetric sawtooth waveform. In a typical system employing a 400 channel multichannel analyzer as a data-storing device, a square wave having its leading edges at channel 0 and 200 can be extracted from the address portion of the multichannel analyzer (Fig. 3.2a). This waveform is then integrated to give a symmetric sawtooth (Fig. 3.2b), ready to be fed into the vibrator amplifier. Feedback circuitry is needed to get the vibrator to follow the waveform precisely. The amplitude of the sawtooth waveform determines the velocity scan of the vibrator. While the vibrator is sweeping continuously over this range of velocities, the multichannel analyzer is sweeping continuously through the 400 channels (Fig. 3.2c) with typical dwell times per channel being about 50 microseconds. The above two sweeps are automatically synchronized and gamma rays from a given velocity are always fed to the same channel in the analyzer. Since each channel corresponds to an equal velocity increment and the dwell times per channel are identical, the spectrum with no absorber will be a flat linear baseline* (Fig. 3.2d). With this symmetric waveform, two mirror image spectra are obtained (Fig. 3.2e), since by the nature of the sawtooth, the velocity of the source will change from plus to minus and then from minus to plus during a single cycle.

This type of electromechanical drive has excellent long term stability and synchronization is ensured in a very simple manner. The velocity range needed for the isotopes that we will consider can easily be obtained with such devices, and the linearity of such systems is typically better than one per cent. The duplicate spectra obtained from such a system are useful because they can easily be folded over (mirror centre about channel 200) and computer averaged. Any lack of linearity in the velocity scan will in general destroy the mirror symmetry, and any spurious absorptions due to electronic faults can be recognized immediately.

Other types of electromechanical drive systems employing sinusoidal or asymmetric sawtooth waveforms have also been used successfully, with perhaps a little more effort. The sinusoidal drive has the disadvantage of a non-linear velocity sweep while rather more elaborate circuits are needed to provide an asymmetric sawtooth that the vibrator can follow and that gives accurate Lorentzian line shapes.

Sources. For a given isotope, there are a number of criteria which are important in choosing a host matrix (ref. 3.2). First, the source should give

---

* The baseline will be truly linear only when the absorber is moved. If the source is vibrated, a slight sinusoidal baseline will result due to the slightly different count rates at the various source positions. However, if the source movement is very small in comparison to the source-detector distance (as it usually is for ^{57}Fe Mössbauer), then any sinisoidal effect will usually be extremely small.

the narrowest possible Lorentzian line to ensure the best resolution. It is essential that each source atom is present in precisely equivalent lattice positions having cubic symmetry. Second, the source should have as large an f factor as possible so that good absorption is obtained. In general, the more ionic or metallic lattices give the largest f factors. Thirdly, the source material should be chemically inert so that the chemical composition does not change by oxidation or hydration. Fourth, the host matrix should not give rise to interfering X-rays, and Compton scattering and photoelectric processes should be minimized. Using a single channel analyzer, we want to be able to select the gamma ray of interest with a minimum of interference from other radiations which will only decrease both the signal to noise ratio, and the resultant quality of the Mössbauer spectrum.

For iron Mössbauer work, the source ^{57}Co is evaporated or electroplated onto a metallic matrix, and the source is then annealed in vacuo at high temperatures. The most common hosts have been metallic Cr, Cu, Pd, and stainless steel. Palladium sources seem to be the most suitable. They give good Lorentzian lines and the line widths approach the theoretical minimum. They are relatively inert chemically, and when prepared properly they give little interference from other radiations, although a 6 keV X-ray has sometimes proved to be troublesome. Stainless steel and chromium matrices give much broader lines while the Cu matrix is more unstable chemically.

For tin Mössbauer, the source (metastable ^{119}Sn) is conveniently produced by the (n, γ) reaction on ^{118}Sn in the thermal flux of a nuclear reactor. A number of sources have been used, including SnO_2, Mg_2Sn, $BaSnO_3$ and Sn diffused into a palladium matrix. SnO_2 gives very broad lines, while Mg_2Sn has a low f factor at room temperature and is rather moisture sensitive. Sn in Pd and $BaSnO_3$ both seem to have very suitable properties for tin sources.

Comparatively little chemical or mineralogical work has been done on other isotopes, and the examination of source properties has correspondingly been rather neglected. In reference 3.2, other source preparations are discussed.

Detectors and electronics. It is desirable to detect the Mössbauer γ ray as efficiently as possible, while discriminating against the remainder of the γ rays, and other radiations such as X-rays. The equipment needed includes a detector, such as a NaI(Tl) scintillation counter, proportional counter or Li drifted Ge counter, a preamplifier, amplifier, a discriminator or single channel analyzer, and a multichannel analyzer. Most of this equipment is available commercially, but in the next few paragraphs we will look at specific features of the above equipment which are important for the Mössbauer spectroscopist.

All three of the above detectors have very high efficiencies for the range of energies required (~ 10 to ~ 100 keV), although the efficiency of the Li drifted Ge detector falls off quite rapidly above 60 keV. The Ge detectors give much better resolution at all energies. A typical 8 cm^2 premium grade detector has a resolution at 50 keV of about 1 keV, while a scintillation counter will give a resolution at 50 keV of typically 10 keV. For isotopes such as ^{182}W where the Mössbauer gamma ray strongly overlaps other radiations, the Ge detectors should lead to much better spectra. However, these detectors normally cost at least ten times that of a proportional or scintillation counter, and they must be kept at liquid N_2 temperatures. In addition, special preamplifiers, cooled to liquid N_2 temperatures, are normally essential for best resolution.

Fortunately, for isotopes such as ^{57}Fe or ^{119}Sn, relatively good resolution of the Mössbauer gamma ray can be obtained with the cheaper scintillation or proportional counters. Using very thin (<1 mm) NaI(Tl) crystals, the 14·4 keV gamma ray can be resolved adequately. Thicker crystals are used for higher energy gamma rays. Integral detector assemblies including the crystal and preamplifier are now very widely used. Below 20 keV, the gas proportional counters have markedly better resolution and a better signal to noise ratio than scintillation counters. Proportional counters are normally filled with Xe, Kr, or Ar, with methane or nitrogen as a quenching gas.

Proportional counters need very high voltage supplies (up to 3000 volts) and stable low capacitance preamplifiers. Scintillation detectors often operate at voltages around 1000 volts and charge-sensitive preamplifiers are normally desirable. The germanium detectors have to be mounted especially to minimize stray capacitance and very low noise charge sensitive preamplifiers are essential for maximum resolution.

For convenience and best results, it is essential to use a multichannel analyzer in the time or multiscaling mode of operation. The dwell time per channel is normally crystal controlled and very precisely regulated. Dwell times per channel are adjustable and normally will vary from about 10 μs to 100 μs. The 'dead time' per channel is normally between 5 μs and 15 μs. It is of course desirable to have an analyzer with as small a dead time as possible so that the maximum count rate is achieved. How many channels are needed for the best results? The theoretical resolution increases as the number of channels increases. However, the count rate drops off proportionately and the linearity of the drive system may not be good enough to justify the increase in channels. For iron spectra, 400 channels seem to be adequate, although many laboratories are now using bigger analyzers for complex spectra of iron containing minerals and other elements.

Finally, it seems important to mention another problem associated with counting. Many minerals contain very small percentages of iron, or other Mössbauer isotopes. Using a 10 mCi ^{57}Co source, satisfactory spectra cannot be obtained in a few days for samples containing less than 1 wgt. per cent iron. Increasing the source strength or moving the source too close to the

Mössbauer spectroscopy

detector decreases the Mössbauer absorption, because typical amplifiers and single channel analyzers become very inefficient at 10–20 KHz rates (10 000 to 20 000 counts s^{-1}) and useless above this. Multichannel analyzers, however, are capable of storing data at much higher rates.

Systems using proportional counters and fast low-noise amplifiers have now been developed to give excellent resolution of the 14 keV gamma ray and to increase the count rate appreciably. Using such counting systems, a 50 mCi ^{57}Co source, and absorbers containing less than one per cent Fe, satisfactory spectra can be readily obtained.

3.2 Other experimental methods

Absorbers. A uniform absorber with randomly oriented crystallites can usually be prepared very easily by sandwiching the finely ground compound or mineral between two layers of cellotape in a Pb holder. If only a small amount of sample is available, and complete recovery is essential, it is best to make a special perspex holder with thin windows. The sample is often mixed with an inert matrix such as finely ground perspex or aluminium. For best line shape and resolution it is desirable that the absorber is 'thin'. As discussed in the previous chapter (eq. 2.25), the line width increases as the 'thickness' of the sample increases.

i.e. $$\Gamma_{ex} = \Gamma_a + \Gamma_s + nK \qquad (3.1)$$

where n is in mg cm^{-2} and K is a constant. For iron Mössbauer, if 10 mg cm^{-2} of iron is used in an absorber, $nK \ll \Gamma_a + \Gamma_s$.

If heavy elements are present in the sample, the per cent absorption often drops off noticeably. Some compounds containing heavy elements will scatter or absorb the Mössbauer gamma ray, and/or emit photons of similar energy which are detected along with the Mössbauer photon. The signal to noise ratio will, of course, decrease. In these cases, thinner absorbers invariably give better absorption.

Unless special care is taken, platey samples such as micas invariably give oriented samples, even when diluted with an inert matrix. The resulting asymmetry of the quadrupole doublets usually contributes to difficulties in the computing procedure. For example, it is no longer possible to constrain the areas of the component peaks to be equal. The following procedure has been found to give truly random samples. The finely ground mineral (>100 mesh) is mixed with a finely ground inert matrix such as perspex. The uniform mixture is then dropped through a fine tipped pipette into the sample holder. The absorber holder should not be pressed or touched (ref. 3.3). Grinding minerals with three to four times the volume of sucrose under acetone gives a floculent powder which also prevents orientation (ref. 3.4).

On the other hand, if oriented absorbers are desired, and large cleavage flakes are not available, an oriented sample may be prepared by allowing a

Experimental techniques

finely ground sample to settle onto a perspex disc through a large column of acetone or other liquid. Because perspex is slightly soluble in acetone, the oriented mineral will remain stuck to the disc.

The cosine effect and source to detector distance. The source to detector distance should be chosen so that the maximum count rate and absorption is obtained and the so-called cosine smearing effect is minimized. The first of these has already been discussed in connection with overloading the electronics at high count rates. The second effect, which both broadens the line and shifts it, can be minimized by collimating the γ beam. Consider a source having a radius R_S at a distance L from a detector having a radius R_D. The emitted gamma photons will make an angle θ with the direction of relative motion (Fig. 3.3a). The energy shift of the γ ray due to the Döppler motion is then

$$\Delta E = (v_0/c)E_\gamma \cos \theta. \quad (3.2)$$

The effect of the $\cos\theta$ term is illustrated in Figure 3.3b for $R_D = R_S$ and where $\alpha = R_D/L$. The ordinate represents the fraction of the total number of photons detected at a velocity v, with a relative velocity v_0. If α is large, an appreciable percentage of the counts are recorded at velocities different from v_0 and broad shifted lines will result. In practice it is desirable to have $\alpha < 0.05$, but it may be necessary to increase α to obtain reasonable count rates with very weak sources.

Calibration. As in all forms of spectroscopy, it is essential to have a method of accurately measuring the total energy scan of the source radiation, and also to have a standard reference line position against which other positions may be quoted. The latter point is especially important, since the different sources used (e.g. ^{57}Co in Pd, Cr, Cu) all emit gamma rays of slightly different energy.

The six line positions from the spectrum of a natural iron foil have been accurately measured and widely used to measure the total velocity scan of a vibrator. A graph of the known iron velocities versus the peak channel numbers is then plotted (Fig. 3.4), and a straight line fitted by the method of least squares. The slope of the line gives the velocity increment per channel. Usually only the inner four lines are used for calibration since this velocity scan is adequate for the spectra of all paramagnetic substances and the outer line positions can be in error by as much as four per cent due to saturation effects. However, the centroid of the iron spectrum has been found to vary slightly from foil to foil, and thus another reference velocity has been sought. The centroid of the two line sodium nitroprusside spectrum is now widely used for this purpose. (See Appendix 2 for conversion velocities.) These two peak positions have also been used to measure the velocity scan. However, iron seems to be better for the latter purpose, since the nitroprusside spectrum

Mössbauer spectroscopy

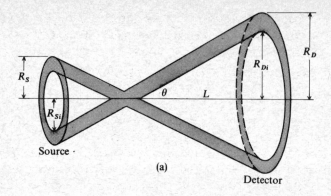

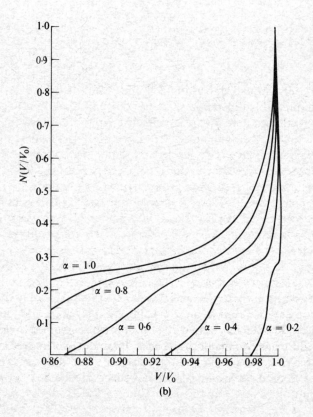

Fig. 3.3 The cosine smearing effect (from J. J. Spijkerman, F. C. Ruegg and J. R. DeVoe, *Mössbauer Effect Methodology*, vol. 1, ed. I. J. Gruverman, Plenum Press (1965)). The parameter α is the aperture (R_D/L) in the case of identical source and detector radius $(R_S = R_D)$.

Experimental techniques

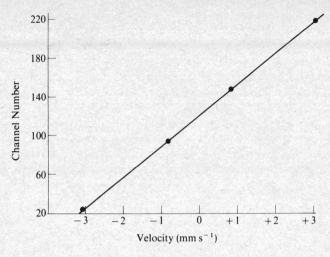

Fig. 3.4 Calibration plot of velocity versus channel number for the inner four peaks of iron metal. The slope of this line gives the velocity increment per channel (0·0316 mm s^{-1} per channel in this case).

consists of only a quadrupole doublet, and several workers have recently obtained appreciably different results for this quadrupole splitting.

It is interesting to note that if a symmetric sawtooth waveform is used to drive the vibrator, the computed mirror or scan centre can be an excellent velocity standard relative to the source used. For example, if the scan centre of a 400 channel instrument is channel 202, the points of zero velocity (relative to the source) are at channels 102 and 302 (assuming a symmetric and linear waveform). Since we have found that the scan centre varies more than the velocity increment per channel, it is certainly advantageous to refer all centre shifts to this scan centre, and then convert to the desired reference standard.

Although natural Fe and sodium nitroprusside appear to be very good calibration substances for iron Mössbauer, considerable calibration problems often arise when other isotopes are being used. Usually, the source in question is replaced by a ^{57}Co source, an iron calibration is taken, and the source is then replaced. The velocity scan with both sources is assumed to be identical. This procedure has two major drawbacks. First, the two sources are usually not the same size and weight, and there may be slight differences in velocity calibration. Second, the iron peaks do not cover an adequate velocity range to be a good calibrant for many other isotopes. The six Fe peaks cover the range $\pm 5\cdot 3$ mm s^{-1} and for isotopes such as ^{121}Sb and ^{129}I, a velocity scan of a few cm s^{-1} is required. For such a velocity scan, the Fe peak positions cannot be accurately estimated, and the velocity scan outside these iron peaks cannot be assumed to be linear as is usually done.

Mössbauer spectroscopy

It is highly desirable then to have a different method of calibration for other Mössbauer isotopes, but reference compounds having accurately measured and stable peak positions covering the appropriate velocity range are often not available. The laser interferometer method briefly described here appears to provide an excellent calibration method for isotopes other than ^{57}Fe. A block diagram of the equipment is shown in Fig. 3.5. A mirror is attached to one end of the vibrating rod and the source is attached to the other. A coherent laser beam (e.g. He–Ne, $\lambda = 6328 \cdot 198$ Å) is split into two, one half going to the stationary mirror B, and being reflected (path 1), and the other half travelling to the vibrator mirror and being reflected (path 2). The two reflected beams are recombined at A. By varying

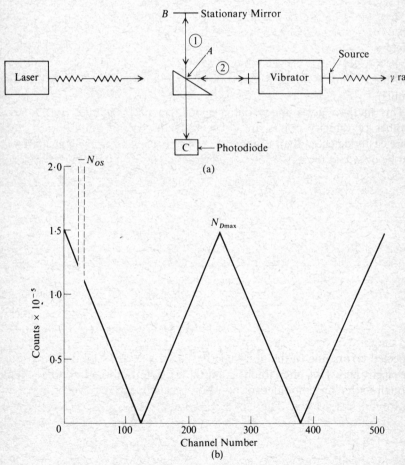

Fig. 3.5 The laser interferometer method of calibration. (a) block diagram of the laser interferometer, (b) resultant multichannel plot of the fringe rate. The number of counts in each channel measures the absolute velocity in that channel (see text).

the path length of beam 2, the two beams interfere constructively or destructively, and the resultant beam is detected at C. For each $\lambda/2$ of motion, constructive interference takes place and one fringe results which gives two pulses at C. For 1 mm motion of the vibrator, 6320·914 counts (6328·198 × 0·25 × $10^{-7})^{-1}$ are detected. The total distance travelled (in mm) is given by $N_D/6320·914$, where N_D is the number of counts detected. Since velocity = distance/time, we must have an accurate method of recording the time. Crystal oscillator pulses (at, for example, 50 000 counts s^{-1}) are recorded while the vibrator is moving. The time in seconds is then given by $N_{OS}/$50 000, where N_{OS} is the total number of oscillator pulses recorded.

The velocity is now given by distance/time, 50 000 $N_D/6320·914\,N_{OS}$. In practice, the oscillator pulse is stored in the first few channels of the memory, and the interferometer pulses are multiscaled. Figure 3.5b shows a typical multichannel plot using a symmetric sawtooth waveform. Zero velocity corresponds to the two minima (no interference fringes counted) while the three maxima correspond to $-V_{max}$, $+V_{max}$ and $-V_{max}$ (V_{max} = 5·35 mm s^{-1} in Fig. 3.5b; N_{OS} = 2·17 × 10^5 counts, $N_{D_{max}}$ = 1·47 × 10^5 counts).

This method gives an absolute velocity in each channel, and a large number of velocity points are collected enabling any nonlinearity to be more easily detected than using an Fe calibration. The equivalent spectrum can be obtained for any velocity range of the vibrator. This calibration can be taken in five to ten minutes, or a selected number of velocity points can be interlaced with a Mössbauer spectrum during its accumulation. The latter method minimizes the effect of any velocity drifts from spectra to spectra.

A good spectrum. How long should a spectrum be counted for, and when can we say that a spectrum is satisfactory? Since the standard deviation σ of a count N equals $\sqrt{N}$, it is obvious that the spectral quality improves as the number of counts increases. For example, if 1 × 10^5 baseline counts (B) are recorded, σ is about 316, or $\sigma/B \sim 0·3\%$; if 1 × 10^6 baseline counts are recorded, $\sigma/B = 0·1\%$. Normally, a spectrum should be counted until the standard deviations in the parameters are substantially smaller than the expected errors due to the non-linearity of the velocity scan or any drift in the electronics. This often implies that the spectrum should be counted long enough so that any shoulders or non-Lorentzian line shapes will be apparent on careful observation. The length of counting may, however, be dictated by other factors such as the cost and strength of sources, and the availability of machine time.

Background. For the best spectra, it is important to maximize the signal to noise ratio. The choice of a counter having the best resolution is usually most important for selecting the Mössbauer γ ray, but other measures may be taken to improve the signal to noise ratio. Unless f factors are being measured, it is not essential to know the exact signal to noise ratio, and fortunately for most mineralogical and chemical applications, the f factor is not needed.

There are different types of background which may be counted along with the Mössbauer gamma ray: natural room background and other radiations such as Compton spectra from higher energy γ rays, other γ rays, and X-rays. The natural background may be reduced by shielding the counter with Pb, but the problem of reducing background from other radiations is much more difficult. If the unwanted radiation is slightly higher in energy than the Mössbauer transition, a critical absorber whose K absorption edge lies between the two radiation energies will be sufficient. The position and thickness of this absorber is critical. For example, in Sn Mössbauer, a Pd foil absorbs the 25·3 keV X-ray of ^{119}Sn, thus enabling the 23·8 keV Mössbauer line of Sn to be detected more efficiently. However, the Pd foil in turn gives 21·2 keV X-rays, but by varying the thickness and position of the foil, the number of these X-rays entering the counter can be minimized. For cases in which the background radiation is lower in energy than the Mössbauer energy, absorbers (usually of lower atomic number than the Mössbauer isotope) may improve the signal to noise ratio. Coincidence techniques may also enhance the signal to noise ratio, but these normally decrease the count rate drastically.

3.3 Computational methods

In all forms of spectroscopy, it is desirable to obtain the most precise and accurate estimates of peak positions, and often of peak widths and areas. Peak positions are often used as a 'fingerprint' to characterize a particular atom or groups of atoms, and the areas under the peaks are often proportional to the number of these atoms or groups of atoms.

In order to obtain the most precise estimates of peak parameters in Mössbauer spectroscopy, it is essential to compute the spectra. For computing to be worthwhile, it is desirable that, (a) the line shapes be known precisely, (b) the variance or standard deviation of a particular count be a known and well-defined function of the number of counts, (c) the no absorption region (baseline) be a linear or well defined function of velocity and, (d) the spectral output be available in digital form, ready to be punched onto computer tape. Fortunately, Mössbauer spectra meet these conditions almost ideally and the area, width and position of a peak along with their standard deviations can be readily obtained. For example, the line shape of a peak is essentially an ideal Lorentzian, the deviation of a count N (using standard counting equipment and a multichannel analyzer in the time mode) is equal to $\sqrt{N}$, the baseline is a linear function of velocity if a sawtooth drive waveform is used (see footnote to page 49), and the information is readily available from multichannel analyzers in digital form.

Computational methods are especially important for Mössbauer spectra of minerals because of their usual complexity and also because of the need to obtain accurate areas for many spectra. In fact, for many spectra it is not at all immediately obvious how many peaks are present. Some spectra contain overlapping lines which visually appear to be almost Lorentzians. The general method used to fit these spectra will be briefly described. Excellent computer programs, such as the one devised by A. J. Stone at Cambridge, are now readily available on request (ref. 3.5).

The method. If a 400 channel multichannel analyzer is used, the observed spectrum consists of 400 numbers defining an envelope. The problem thus consists of finding the positions, widths and intensities of a number of Lorentzian lines which, when superimposed, give the best fit to the observed envelope. The line shape, $Y(X)$, of the envelope for a Lorentzian line is given by:

$$Y(X) = b - \frac{Y(0)}{1 + \left(\frac{X - X(0)}{\Gamma_{ex}/2}\right)^2} \quad (3.3)$$

where $Y(0)$ is the intensity at the resonance energy (or velocity) $X(0)$, Γ_{ex} is the width at half height, and b is the baseline intensity. Thus the equation of the envelope for n lines becomes:

$$Y(X) = b - \sum_{i=1}^{n} \frac{Y(0)_i}{1 + \left(\frac{X - X(0)_i}{\Gamma_{ex}/2}\right)^2} \quad (3.4)$$

Finally, a small baseline term can be applied to correct for a slight sinusoidal variation of the baseline due to source movement. The magnitude of this term will depend on the source–detector distance and the source movement.

For an n line mirror image spectrum, $Y(X)$ is a function of $3n + 3$ parameters (position, width, and intensity of each line, plus the baseline, baseline correction and scan centre) denoted q_i and written as a vector $\mathbf{q}$. We want to minimize χ^2, which is analogous to the X-ray R factor. The sum of the squared deviation from the best fit divided by the variance at a single count (χ^2), can be expressed as:

$$\chi^2 = \sum_{r=1}^{400} W_r[Y_r - Y(X_r|\mathbf{q})]^2 \quad (3.5)$$

where Y_r is the observed count at channel r, $Y(X_r|\mathbf{q})$ is the function (3.4) above and W_r is the inverse of the variance for channel r. Initial estimates are chosen for the $3n + 3$ parameters q_i, and using

$$\frac{d\chi^2}{dq_i} = 0 \qquad (3.6)$$

for each q_i, corrections are determined for each q_i such that χ^2 is minimized. This constitutes one iteration. The procedure is then repeated using the corrected estimates from the previous iteration, until the values of χ^2 obtained in successive iterations differ by less than a suitable small quantity, chosen usually to be 10^{-6}. This normally takes about 10 iterations, but the number of iterations varies greatly depending on the complexity of the spectra and the accuracy of the initial estimates.

For complex spectra, it is often essential to hold parameters constant or constrain a number of parameters to be equal throughout all, or any part of the fitting procedure. Without these constraints, it is often not possible to get a complex spectrum to converge. Many programs enable these constraints to be applied very simply. Examples of the use of constraints will be outlined in a later section.

Criteria for a 'good fit'. How can one be confident that the correct or 'best' fit to a spectrum is obtained? This is often a difficult problem, but there are three main questions which should be asked, taking iron containing minerals as an example.
1. Is the χ^2 value statistically acceptable, or failing this, is the χ^2 value similar to those of simple spectra of similar quality?
2. Are the number of sites in which iron is found consistent with what is known from other evidence, such as X-ray or infrared?
3. Are the Mössbauer parameters for an iron atom consistent with previous Mössbauer work on similar coordination sites?

It is thus desirable to have related information from other techniques on a particular mineral, but the χ^2 value is still a very useful indication of the quality of the fit.

For a fit to be statistically acceptable, it is required that the χ^2 value be between the 1% and 99% points of the χ^2 distribution; that is between $(v + 2 \cdot 2 - 3 \cdot 3\sqrt{v})$ and $(v + 2 \cdot 2 + 3 \cdot 3\sqrt{v})$ where: v, the number of degrees of freedom is the number of channels used in fitting the spectrum less the number of adjustable parameters in the fitted curve. The theoretical statistically acceptable upper limit is normally between 430 and 450 for a 400 channel spectrum. In order to obtain a statistically acceptable χ^2, the spectrometer must be working perfectly. Firstly, the two halves of a spectrum must be perfect mirror images (if the spectra are folded over during computation) requiring that the velocity coil waveform be a perfectly linear and symmetric sawtooth. Secondly, the variance of a count can be modified by dead time effects in the counting equipment. Thirdly, multichannel analyzer faults can lead to spurious counts and high χ^2 values. Of these, only the first appears to be normally important. Very large χ^2 values are obtained even when very small deviations from mirror symmetry (~ 1 channel) are present.

Non-instrumental causes of poor χ^2 values should also be mentioned. Firstly, it will often be impossible to resolve contributions to the spectra from an iron atom, if it contributes less than five per cent of the total intensity of the spectrum. This small amount of unresolved iron could give rise to a χ^2 which is at least 50 too large. Secondly, the line shapes may not be the assumed Lorentzian shape. Non-Lorentzian line shapes could be caused by poorly prepared sources, very 'thick' absorbers, or the strong overlap of a number of Lorentzians which are not resolvable.

Considering the second question, it is obvious that if a mineral has three cation sites (from an X-ray study) and all the iron is present as Fe^{2+} (from chemical analysis), the maximum number of peaks in the spectrum should be six (three doublets). A number of these peaks may be so closely overlapping that they cannot be resolved no matter what reasonable constraints are employed (see the anthophyllite spectrum in the next section). On the other hand, a six-peak fit may give large χ^2 values or unreasonable parameters, in which case it is possible that the chemical analysis is in error (e.g. Fe^{3+} is present) or that Fe^{2+} is present in more than three sites. More peaks should then be fitted to see if the χ^2 value drops appreciably and if the parameters become more reasonable with those expected from previous experience. An example of such spectra is given by the omphacite spectrum in the next section. Despite some inconsistencies between Mössbauer and other techniques, it is always desirable to have other information. For example, without the X-ray structure and chemical analysis, our overall confidence in a particular fit certainly decreases.

Looking at the third question, the centre shifts and quadrupole splittings for Fe^{2+} and Fe^{3+} in silicate minerals lie within fairly well defined regions. For example, six-coordinate Fe^{2+} in silicate minerals gives centre shifts between 1·30 and 1·43 mm s^{-1} with respect to sodium nitroprusside. A fit which gives a six-coordinate Fe^{2+} centre shift outside this region should be regarded with some suspicion. Line widths are often extremely helpful in indicating the number of distinct (to Mössbauer) iron sites. Full widths at half height for Fe^{2+} in silicate minerals invariably fall between 0·28 and 0·35 mm s^{-1} using a ^{57}Co in Pd source. If a line is much broader than 0·35 mm s^{-1} (say 0·50 mm s^{-1}) and/or if there is any hint of asymmetry, it is quite likely that there are two or more overlapping but resolvable lines present. If χ^2 decreases markedly (>50) on fitting two lines, then this is evidence for the existence of appreciable amounts of iron in two sites.

It is often useful to run complex spectra at a number of different temperatures, because the resolution can often change markedly. The temperature dependence of the quadrupole splittings are especially sensitive to slight differences in site distortions, and a line or lines which are unresolved at 295 K may be resolvable at a higher or lower temperature.

In addition, for unravelling very complex spectra, it is often useful to take spectra of a number of minerals of the same type. The centre shifts, quadrupole splittings, and line widths should be similar for a group of minerals such as P2 omphacites.

Examples of the fitting procedure. It is essential, for complex spectra, to proceed by fitting initially the minimum number of lines with many constraints to ensure that the fit will converge. After obtaining the best parameters from an incomplete fit, it is then usually possible to proceed to the next stage with much better initial estimates for the parameters. The constraints can then normally be released in later stages of fitting.

The fitting process will now be illustrated for two mineral spectra: an anthophyllite containing 23.0% Fe^{2+}, and an omphacite having P2 symmetry. It is hoped that these two examples will be a useful guide for fitting any complex spectra of chemicals or minerals.

For the purposes of this section, it is not necessary to know the detailed mineralogy of these minerals. It is sufficient to know that there are a maximum of four cation sites for the iron in the anthophyllite, and eight cation sites (two pairs of four) in the omphacite.

Only two narrow peaks characteristic of Fe^{2+} are immediately apparent in the anthophyllite spectrum (Fig. 3.6a). An initial two-peak fit constraining the widths of the lines to be equal gives a χ^2 of well over 500, but the line widths (0.34 mm s^{-1}) and other parameters are reasonable for Fe^{2+} in six-coordination. The rather large χ^2 and the four possible sites for the iron should prompt one to look more carefully at the spectrum. On the outside of the two main peaks there is an indication of shoulders. A four-peak fit constraining the half widths in pairs converged and gave a χ^2 of 435. In the next stage of fitting the half width constraint was removed, and χ^2 decreased to 431. This fit is illustrated in Fig. 3.6b. The half widths in this fit are all under 0.35 mm s^{-1}, and attempts to fit more than four peaks failed. We conclude that we can only resolve Fe^{2+} in two sites, implying that there is either Fe^{2+} in only two sites, or an accidental overlap of lines from appreciable quantities of Fe^{2+} in three or two sites. The spectrum of an anthophyllite containing a larger amount of ferrous iron (Fig. 3.6c) gives better resolution of the outer doublet. The parameters for both anthophyllites are identical within statistical error.

In the anthophyllite spectrum, half width constraints were employed in the initial fits. Are these and other constraints justified, and can they lead to misleading or erroneous answers? The half widths of the component peaks of Fe^{2+} quadrupole doublets have been found to be very closely equal for

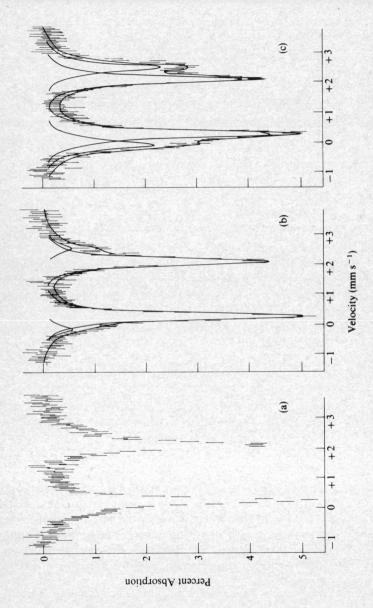

Fig. 3.6 Mössbauer spectra of anthophyllites. (a) Computer plot of raw data for 23·0% Fe^{2+}. (b) Four peak fit to spectrum (a). (c) Four peak fit to 31·6% Fe^{2+} anthophyllite. [G. M. Bancroft, A. G. Maddock, R. G. Burns and R. G. J. Strens, *Nature*, **212**, 913 (1966).]

a wide variety of chemical and mineralogical samples. This constraint then is certainly justified on the basis of previous experience. For Fe^{3+} the line widths may be appreciably different, so that this constraint should be employed with caution for Fe^{3+} peaks. For a randomly oriented sample, the areas of the component doublet peaks are always very close to 1:1, and constraining the areas to be equal is certainly justified. Causes of deviation from a 1:1 ratio have been mentioned in the last chapter.

For very complex spectra, such as the omphacite spectrum described shortly, it is necessary to employ further constraints to get the fit to converge. The half widths for most Fe^{2+} peaks in silicate minerals are 0.32 ± 0.03 mm s^{-1}, and thus we feel that constraining all the half widths of Fe^{2+} peaks to be equal is usually justified, although the resulting areas will be less accurate. It should be emphasized that although line position constraints may be used for the initial stages of the fitting process, any line position constraints in final fits may well lead to misleading results.

We turn now to the omphacite spectrum (Figure 3.7a). This consists of two rather broad envelopes with very few distinctive features. The high velocity envelope is due to Fe^{2+} (since Fe^{3+} components never occur at this velocity), while the low velocity envelope is due to the corresponding Fe^{2+} quadrupole pairs along with any Fe^{3+} peaks. As mentioned earlier, there are eight appropriate cation sites for Fe^{2+} and Fe^{3+}, so that if both Fe^{2+} and Fe^{3+} enter all eight sites, there is a possibility of thirty-two peaks. It has never been possible to resolve the various Fe^{3+} doublets from different six-coordinate sites, so that we would expect Fe^{3+} in any or all of the eight sites to give rise to just one doublet. This leaves us with eighteen possible peaks, sixteen from Fe^{2+}, and two from Fe^{3+}. It is highly unlikely that the sixteen ferrous peaks would be resolved, and we might expect either four or eight Fe^{2+} peaks, each peak being an unresolved overlap of four or two peaks respectively.

We begin by fitting this spectrum to six peaks (Fig. 3.7b). There are obviously two high velocity Fe^{2+} peaks (A and B) which implies that there are the two corresponding peaks A' and B' under the low velocity envelope. The centre shifts for these two doublets are estimated at a similar value to those from other six-coordinate sites—about 1.4 mm s^{-1}. The greater intensity of the low velocity envelope is due to Fe^{3+} peaks, and since Fe^{3+} normally gives a doublet, two peaks, C and C' are fitted. These two peaks must lie on the high velocity side of the low velocity envelope to obtain a reasonable centre shift value. The areas of the peaks are constrained in pairs.

There are several pieces of evidence which suggest that this fit is incomplete. First, the half widths of the peaks are very dissimilar, e.g., the width of peaks A and A' is about 0.33 mm s^{-1}, while the width of peaks B and B' is about 0.55 mm s^{-1}. Second, the χ^2 value is about 450, slightly larger than

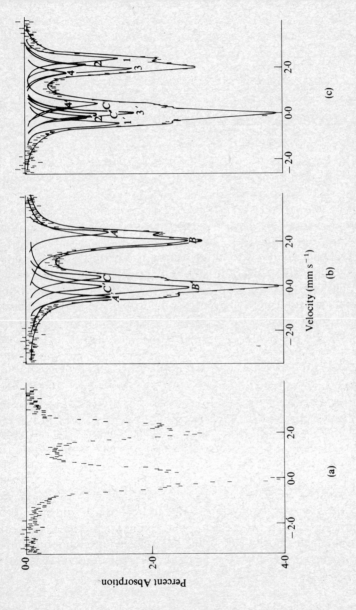

Fig. 3.7 Mössbauer spectra of an omphacite. (a) Computer plot of raw data. (b) Six-peak fit. (c) Ten-peak fit. [G. M. Bancroft, P. G. L. Williams and E. J. Essene. *Mineral Soc. Amer. Spec. Pap.*, **2**, 59 (1969).]

usual. Third, the six-peak fit to the spectrum at 80 K is inconsistent with a six-peak fit at room temperature.

We then fit eight peaks, constraining just the areas in pairs. This fit converged, and χ^2 decreased by about 50. At this stage, the χ^2 value of about 400 was quite reasonable, and on this basis alone, we may have stopped at this point. However, the half widths were still markedly different from peak to peak and it is highly unlikely from the X-ray structure that there is Fe^{2+} in three or six sites as opposed to two, four, or eight sites. It appeared, then, highly probable that there were indeed four ferrous doublets, but to fit ten peaks to these spectra required many constraints in order that the fitting process did not diverge. We constrained all the low velocity peaks as in the eight line fits, and proceeded to fit four peaks (1, 2, 3, 4) in the high velocity region, constraining the half widths and intensities of these peaks to be equal to each other, but not at any fixed value. Again χ^2 decreased by about 20. The intensity constraints were then released, giving only a marginal decrease in χ^2. We then fit four corresponding peaks (1', 2', 3', 4') in the low velocity region, constraining the intensities in pairs, the half widths at the best value from the previous fit, and attempting to pair the peaks off so that the centre shift values for all four doublets were approximately equal. At this stage, the χ^2 value often increased markedly, but it should be realized that we are now really forcing the fit in order to obtain reasonable parameters. For example, the assumption of equal intensities is not perfect, and the Fe^{3+} lines are due to a superposition of lines from Fe^{3+} in a number of sites. These Fe^{3+} peaks will not be Lorentzian in shape.

This fit (Fig. 3.7c) gives reasonable parameters for six-coordinate Fe^{2+} and Fe^{3+} without constraining any positions. More important, the centre shift, quadrupole splitting, and widths are in good agreement with those from other omphacites. Also, the spectra of omphacites at 80 K are consistent with the above interpretation. The areas of the three unresolved Fe^{2+} peaks cannot of course be determined accurately. All one can say is that appreciable amounts of iron enter more than two, and probably four, distinct sites.

From the above examples, it is obvious that computer fitting is essential to obtain the maximum amount of information from a complex Mössbauer spectrum. It is also obvious that other data from X-ray work or chemical analysis are often essential.

Problems

1. In ^{237}Np Mössbauer spectra (see W. L. Pillinger and J. A. Stone, *Mössbauer Effect Methodology*, Vol. 4, p. 217), velocities in the ± 20 cm s^{-1} range are normally needed to observe all the peaks in a spectrum. The

natural linewidths are less than 0.1 mm s^{-1}. If natural linewidths are observable, can spectra be obtained in the constant acceleration mode using a 400 channel analyzer? How many channels would probably be needed?

2. Using a 512 channel multichannel analyzer, a symmetric sawtooth wave form and a ^{57}Co/Pd source, it was found that the four inner peaks of Fe foil are at channels 22·05, 93·85, 148·14 and 219·97, and the scan centre is 255·24 channels. What is the velocity increment per channel? Take the four Fe lines to be at -3.083 mm s^{-1}, -0.838 mm s^{-1}, $+0.838$ mm s^{-1} and $+3.083$ mm s^{-1} respectively. What is the zero velocity channel with respect to the source, and with respect to the absorber? What is the centre shift of iron with respect to ^{57}Co/Pd?

3. The mineral deerite (approximate composition $Fe^{2+}_{13}Fe^{3+}_{7}Si_{13}O_{44}(OH)_{11}$) gives a complex Mössbauer spectrum (see Figure 7.3 and G. M. Bancroft, R. G. Burns and A. J. Stone, *Geochem. Cosmochim. Acta* **32**, 547 (1968)). Some of the peaks are broad, suggesting that they are a superposition of a number of closely overlapping peaks. χ^2 is 445 for this fit ($\#1$).

Using different initial guesses of the parameters, it is possible to obtain another fit ($\#2$) with $\chi^2 = 469$ with the same parameters for peaks A and A', but substantially different parameters for peaks B,B' and C,C'. Can you be confident that fit $\#1$ is correct? What other experiments would be helpful?

References

3.1 M. Kalvius, in *Mössbauer Effect Methodology*, Plenum Press, I. J. Gruverman, ed., 1965, vol. 1.
3.2 N. Benczer-Koller and R. H. Herber, *Chemical Applications of Mössbauer Spectroscopy*, Academic Press, V. I. Goldanskii and R. H. Herber, eds., 1968.
3.3 M. G. Clark, G. M. Bancroft and A. J. Stone, *J. Chem. Phys.*, **47**, 4250, (1967).
3.4 C. Greaves and R. G. Burns, *American Mineral.*, **56**, 2010, (1971).
3.5 A. J. Stone, appendix to: G. M. Bancroft, A. G. Maddock, W. K. Ong, R. H. Prince and A. J. Stone, *J. Chem. Soc.* (A), 1966, (1967).

Bibliography

T. C. Gibb and N. N. Greenwood, *Mössbauer Spectroscopy*, Chapman and Hall, 1971.
I. J. Gruverman ed., *Mössbauer Effect Methodology*, vol. 1–7, 1965–1972.

4. Mössbauer spectroscopy as a fingerprint technique in inorganic chemistry

We indicated in previous chapters that the Mössbauer centre shift and quadrupole splitting of a useful Mössbauer isotope are sensitive to the electronic structure of the Mössbauer atom, and the ligand environment about that atom. For example, in the case of ^{57}Fe, the centre shift is very sensitive to the formal oxidation state of the iron atom (Table 4.1) and

Table 4.1 Centre shifts (C.S.) for high spin iron compounds relative to nitroprusside* (mm s^{-1})

Oxidation state	+1	+2	+3	+4	+6
C.S.	~ +2.2	~ +1.4	~ +0.7	~ +0.2	~ −0.6

*All centre shift values for Fe compounds in this book are quoted relative to sodium nitroprusside [Na$_2$Fe(CN)$_5$NO].2H$_2$O. (See Appendix 2 for velocity conversion from source data.)

becomes more negative as the formal oxidation state increases. Figure 7.1 also indicates the sensitivity of centre shift and quadrupole splitting to the oxidation state and coordination number of Fe.

Any spectroscopic parameter which is sensitive to the electronic or molecular structure of a compound is capable of providing the chemist or mineralogist with two general types of information: the characterization of the type of atom or molecule by comparison of the spectroscopic parameter with those of known species (the so-called fingerprint technique); and the elucidation of bonding and structure. The first of these applications is very important yet requires very little familiarity with theoretical aspects of the subject. The second application—structure and bonding—is perhaps more important. This information is often difficult to obtain by other methods, but it requires a more thorough grasp of the theory. The first application will be dealt with in this chapter, while the second will be considered in chapters 5 and 6.

The fingerprint application is very useful in the following ways: to indicate the purity of a compound, to characterize the oxidation state of the Mössbauer atom, to prove whether one compound has a different structure from another of the same composition, to indicate whether two or more Mössbauer nuclei in a polynuclear compound are in equivalent environments, and finally to identify the compounds or species in a complex mixture. In this chapter, we will look at several applications of Mössbauer spectroscopy which should demonstrate the usefulness, and limitations of, Mössbauer spectroscopy in the above areas.

Unlike many other spectroscopic experiments, the line shapes in Mössbauer spectra are normally very well-behaved Lorentzians, and as indicated in chapter 3, the spectra can readily be subjected to detailed computer processing. Thus, if there is a shoulder on a peak, or if the peak is slightly asymmetric, one can be confident that there is a peak present which is causing this asymmetry. However, it should be remembered that Mössbauer requires rather large samples (usually > 50 mg), and it will often not be possible to detect a contribution from the spectrum if it is less than five per cent of the area of the main peaks.

4.1 Purity and characterization

As a routine measure, it should be useful for a synthetic inorganic chemist to run a Mössbauer spectrum to indicate the purity of a compound, to characterize the coordination number and oxidation state of the Mössbauer atom, or to prove whether the compound has a different crystal structure from a known compound of the same composition.

A pure compound will normally give narrow Lorentzian Mössbauer lines. Any deviation from Lorentzian shape indicates that impurities are present, and any broadening may indicate that the compound has not been well crystallized so that each Mössbauer atom may not be in exactly the same environment. For example, the spectrum of $H_2Fe(CO)_4$ (Fig. 4.1) shows distinct shoulders on the outside of the main peaks, and the main doublet due to $H_2Fe(CO)_4$ is rather broad ($\Gamma = 0.42$ mm s^{-1}, compared with the normal 0.26 mm s^{-1}). In this case, the broadening is probably due to 'thickness' broadening (eq. 2.25) because a large amount of $H_2Fe(CO)_4$ was inadvertently evaporated onto the cold window. Two peaks, A and A', were fitted in the wings of the large $H_2Fe(CO)_4$ peaks. The centre shift and quadrupole splitting for peaks A and A' were found to be identical to those of $Fe(CO)_5$, which is known to be a decomposition product of $H_2Fe(CO)_4$. The peak areas indicate that the absorber contains about two per cent $Fe(CO)_5$.

Mössbauer spectroscopy

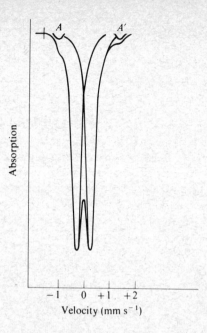

Fig. 4.1 Mössbauer spectrum of $H_2Fe(CO)_4$. Small outside peaks are due to $Fe(CO)_5$. (G. M. Bancroft, M. J. Mays, and B. E. Prater, *J. Chem. Soc.* (A), 956 (1970).)

Mössbauer spectra are often very useful to characterize the oxidation state of the Mössbauer atom, especially in chemically difficult situations. The very different centre shift values for some of the different oxidation states of iron indicate that these should be diagnostic. Any gross deviation of this parameter from the expected value might indicate undesirable reactions. An excellent example which illustrates this point rather well is given by

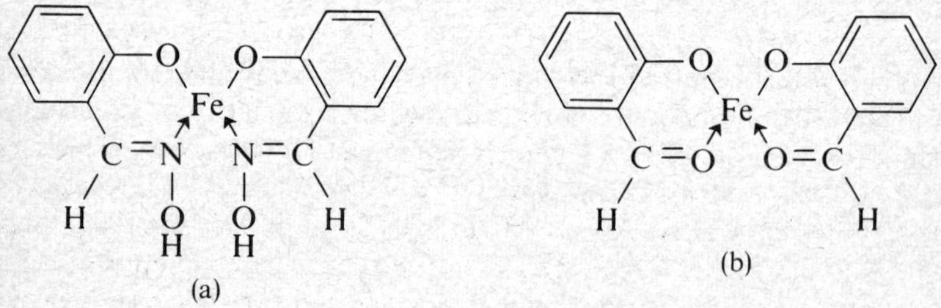

Fig. 4.2 Fe^{II} high spin compounds of (a) salicylaldoxime and (b) salicylaldehyde. (J. L. K. F. de Vries, J. M. Trooster, and E. de Boer, *Chem. Comm.*, 604 (1970).)

the Mössbauer spectra of high spin Fe^{II} chelates of salicylaldoxime, salicylaldehyde, and others (Fig. 4.2). The first reported spectra of these compounds gave centre shift (~ 0.60 mm s^{-1}) and small quadrupole splitting values characteristic of Fe^{III} compounds; but these parameters were still attributed to Fe^{II} compounds, with a large degree of π back bonding. However recent work shows conclusively that the above compounds are in fact oxidation products of Fe^{II} compounds, which give normal Fe^{II} centre shifts and quadrupole splittings of about 1.4 mm s^{-1} and 2.5 mm s^{-1} respectively. Preparation of these Fe^{II} compounds must be carried out in a vacuum system. Mössbauer spectroscopy is very useful here, because the Fe^{II} and Fe^{III} compounds cannot be readily distinguished by magnetic measurements or chemical analyses.

Other Mössbauer isotopes e.g., ^{129}I, ^{121}Sb, ^{119}Sn, ^{129}Xe, ^{125}Te, ^{197}Au, ^{99}Ru, ^{193}Ir, ^{151}Eu, and ^{237}Np have large enough $\delta R/R$ values to give significant changes in centre shift from one oxidation state to another. The centre shift for compounds of the above isotopes has often been used to determine oxidation states—often in chemically difficult situations.

However, there are many situations where Mössbauer spectroscopy cannot readily distinguish between oxidation states. For low spin iron compounds containing π acceptor groups, the centre shifts have a fairly small range for oxidation states from $+3$ to -2. For example, $K_3Fe^{III}(CN)_6$ and $K_4Fe^{II}(CN)_6$ have remarkably similar centre shifts, and CO compounds of different oxidation states also have very similar centre shifts.

For Sn Mössbauer spectroscopy, Sn^{II} and Sn^{IV} compounds normally give characteristic centre shifts. The dividing line for Sn^{II} and Sn^{IV} is normally taken to be about 2.65 mm s^{-1}, with those below being Sn^{IV} and those above being Sn^{II}. However, transition metal compounds such as $Fe^{II}(SnCl_3)_2$-$(ArNC)_4$, which formally contain the $Sn^{II}Cl_3^-$ ligand have centre shift values of about 1.8 mm s^{-1}—in the Sn^{IV} region. Clearly, the centre shift values for such Sn compounds cannot be used to define the oxidation state of Sn. It is probably more reasonable to use the centre shift as a measure of the *valency* of Sn, and all of the $SnCl_3$ compounds contain four-valent Sn.

Finally, another very simple application is to determine whether two apparently similar compounds have different structures. Donaldson showed that a new orthorhombic form of SnF_2 had significantly different Mössbauer parameters from the well-known monoclinic form (Table 4.2). Another example is given by the *trans-* and *cis-*isomers of $FeCl_2(ArNC)_4$ which have quadrupole splittings of $+1.55$ mm s^{-1} and -0.78 mm s^{-1} respectively. This $2:-1$ ratio is rationalized in chapter 6.2. Similarly, compounds such as $Fe^{II}X_2(CO)_2P_2$ (X = Cl, Br, I; P = phosphines, phosphites) have five geometric isomers most of which give significantly different quadrupole splittings. Using the partial quadrupole splitting treatment in chapter 6, these isomers can often be assigned.

Mössbauer spectroscopy

Table 4.2 Mössbauer parameters for orthorhombic and monoclinic SnF_2 (mm s^{-1})

Type	C.S.*	Q.S.
Orthorhombic	+3·30	+2·20
Monoclinic	+3·70	+1·80

* Relative to SnO_2 or $BaSnO_3$, which are within 0·10 mm s^{-1} of each other. All Sn centre shift values are quoted relative to SnO_2 in this book. See Appendix 2 for velocity conversions from other sources. J. D. Donaldson, R. Oteng, and J. B. Senior, *Chem. Comm.*, 618 (1965).

4.2 Detection of structurally different atoms in polynuclear compounds

The Mössbauer effect is often extremely useful for determining whether two or more Mössbauer atoms are equivalent in a polynuclear structure. An outstanding example of this type of application is given by the structure of $Fe_3(CO)_{12}$. Until a definitive crystal structure was published very recently, several different structures were proposed on the basis of the incomplete X-ray work and infrared, either based on a triangle of Fe atoms or linear Fe atoms (Fig. 4.3). The incomplete X-ray work suggested that $Fe_3(CO)_{12}$ had a symmetrical triangular structure with either zero, three or six bridging carbonyls, although the X-ray study did not rule out an unsymmetrical structure.

The Mössbauer spectrum of $Fe_3(CO)_{12}$ (Fig. 4.4) indicates clearly that there is more than one type of iron atom present since the maximum possible number of lines obtainable for a diamagnetic compound at these temperatures is two (chapter 2). This result clearly rules out the symmetrical triangular structures (Fig. 4.3a, b). Because we would expect the centre shift values of all Fe^0 atoms to be reasonably similar, this spectrum must be assigned as follows: outer two lines to one type of iron atom (A), and inner doublet to another type of iron atom (B). Since the area is usually approximately proportional to the number of iron atoms present, it is apparent that the ratio of $Fe_A:Fe_B \simeq 2:1$. Linear structures of the type shown in Fig. 4.3c,d, would be consistent with this spectrum but the spectrum of the $Fe_3(CO)_{11}H^-$ anion, which is very similar to that of $Fe_3(CO)_{12}$ (Table 4.3), rules out a linear structure. If there are any bridging carbonyls in a linear structure, it is not possible to replace any CO group by H and leave two iron atoms equivalent. The above evidence strongly suggests that $Fe_3(CO)_{12}$ has an unsymmetrical triangular structure, with one iron atom coordinated to the same atoms in both $Fe_3(CO)_{12}$ and the hydride, since the parameters for iron atom B in both the parent molecule and the anion are very similar.

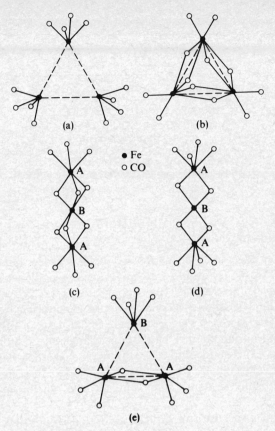

Fig. 4.3 Possible structures for $Fe_3(CO)_{12}$. (After N. E. Erickson and A. W. Fairhall, *Inorg. Chem.*, **4**, 1320 (1965).)

One structure which is consistent with this evidence is given in Fig. 4.3e. The H^- substitutes for a bridging carbonyl, leaving atoms A equivalent, but the centre shift is decreased somewhat from that in $Fe_3(CO)_{12}$. The very similar parameters for Fe_B in both structures is consistent with the identical nearest neighbours for Fe_B in both structures. There are other unsymmetrical triangular structures which would fit the above evidence, but a recent X-ray structure confirms that $Fe_3(CO)_{12}$ has the structure shown in Fig. 4.3e.

Another example of this type is given by the ^{129}I Mössbauer spectrum of $I_2Cl_4Br_2$ (Fig. 4.5). All the peaks are broadened somewhat, and the high velocity peaks (2 and 4) are split into two peaks of equal intensity. The equal intensities make it highly unlikely that the doubling of the spectrum is due to an impurity. Thus the two iodine atoms in $I_2Cl_4Br_2$ are not equivalent. The similarity of the spectrum to that of I_2Cl_6 indicates that $I_2Cl_4Br_2$ has

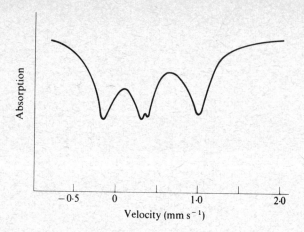

Fig. 4.4 Mössbauer spectrum of $Fe_3(CO)_{12}$. (From R. Greatrex and N. N. Greenwood, *Disc. Farad. Soc.*, **47**, 126 (1969).)

the basic I_2Cl_6 structure (Fig. 4.6a). Structures (b), (c), and (d) in Figure 4.6 are ruled out because the iodine atoms in these structures are equivalent. The centre shift value for one of the iodine atoms is identical to that of I_2Cl_6 indicating that there are four chlorine atoms bonded to one iodine. This evidence would rule out structure (e) and leave us with structure (f) as being consistent with all the above evidence.

Table 4.3 Mössbauer parameters for $Fe_3(CO)_{12}$ and $Na Fe_3(CO)_{11}H$ at 298 K (mm s^{-1})

Compound	Fe atom	C.S.	Q.S.
$Fe_3(CO)_{12}$	A	+0.34	1.05
	B	+0.32	<0.20
$Na[Fe_3(CO)_{11}H]$	A	+0.26	1.32
	B	+0.28	<0.20

N. E. Erickson and A. W. Fairhall, *Inorg. Chem.*, **4**, 1320 (1965).

It should be emphasized, however, that if a polynuclear compound gives rise to apparently one set of absorptions, this negative evidence cannot be taken as proof of equivalence. It is possible that the Mössbauer parameters are just not sensitive to the differences in the ligands on the two or more Mössbauer atoms. Detailed computation then becomes important for detecting any line broadening or asymmetry which could be due to an overlap of Lorentzians.

Mössbauer spectroscopy as a fingerprint technique in inorganic chemistry

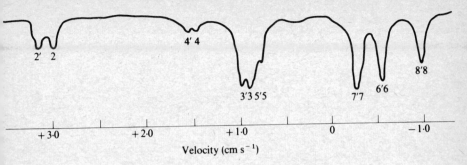

Fig. 4.5 ^{129}I Mössbauer spectrum of $I_2Cl_4Br_2$. (Adapted from, M. Pasternak and T. Sonnino, *J. Chem. Phys.*, **48**, 1997 (1968).)

For example, the complex $(\pi\text{-}C_5H_5)_2Rh_2Fe_2(CO)_8$ apparently gives one doublet with a quadrupole splitting of about 1 mm s^{-1}. However, the line widths (0·36 mm s^{-1}) are appreciably ($\sim$0·10 mm s^{-1}) larger than the widths in analogous compounds, and a four peak fit gives a decrease in χ^2 of about 50. This evidence then strongly suggests that the two iron atoms are not equivalent. Structures such as shown in Fig. 4.7(a) can be ruled out, and structures such as shown in Fig. 4.7(b), (c) would be consistent with the

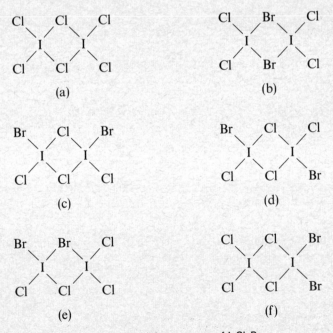

Fig. 4.6 Possible structures of $I_2Cl_4Br_2$.

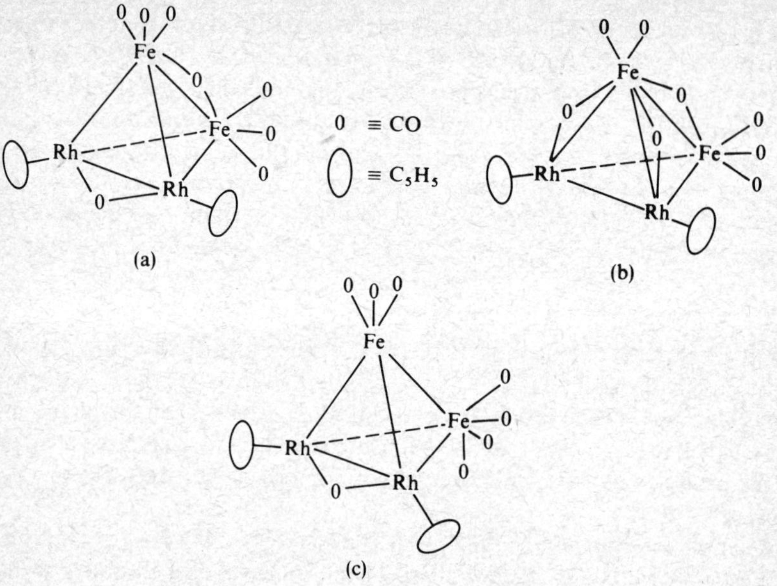

Mössbauer spectroscopy

Fig. 4.7 Possible structures for $(\pi\text{-}C_5H_5)_2Rh_2Fe_2(CO)_8$. (J. Knight and M. J. Mays, *J. Chem. Soc., A*, 654 (1970); M. R. Churchill and M. V. Veidis, *Chem. Comm.*, 529 (1970).)

Mössbauer evidence. A recent X-ray study shows that (c) is the correct structure.

Finally, there are a few polynuclear compounds known that give rise to a single, narrow, quadrupole doublet, yet the iron atoms are not equivalent.

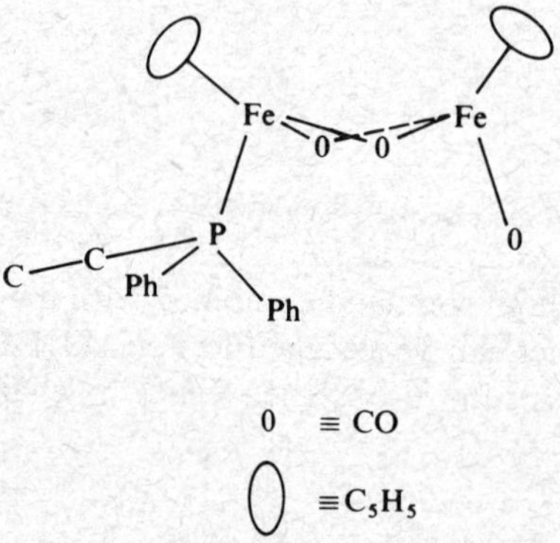

Fig. 4.8 Structure of one half of $[(\pi\text{-}C_5H_5)_2Fe_2(CO)_3]_2\text{DPPA}^*$. (A. J. Carty, T. W. Ng, W. Carter, G. J. Palenik, and T. Birchall, *Chem. Comm.*, 1101 (1969).)
*DPPA $\equiv$ Ph$_2$PCCPPh$_2$.

The crystal structure of $[(\pi\text{-}C_5H_5)_2Fe_2(CO)_3]_2$DPPA shows that the molecule consists of two $Fe_2(CO)_3(\pi\text{-}C_5H_5)_2$ units linked by a DPPA molecule (Fig. 4.8). The two iron atoms are clearly not equivalent, yet the Mössbauer spectrum shows one narrow doublet. For example, the line widths at room temperature (0·24 and 0·26 mm s^{-1}) are the minimum obtainable on the equipment. This result indicates the dangers of inferring equivalence from negative evidence, and suggests that in the above structure, DPPA and CO have a very similar effect on the electronic environment about the iron atom.

4.3 Solid-state decompositions

Mössbauer spectroscopy has been particularly valuable for following reactions in the solid state. The information from Mössbauer spectra can often not be obtained by other techniques. Once again, we are using Mössbauer spectroscopy as a fingerprint technique to assign the structures of the products by comparing observed parameters with those from known species.

A simple example of this type of use is given by the thermal decomposition of $Fe_2(C_2O_4)_3.6H_2O$. The initial spectrum consists of a doublet having a centre shift of 0·64 mm s^{-1} and a small quadrupole splitting (Table 4.4).

Table 4.4 Room temperature Mössbauer parameters (mm s^{-1}) for the thermal decomposition of ferric oxalate, $Fe_2(C_2O_4)_3.6H_2O$

Temperature of heating, °C	C.S.	Q.S.	H KOe	Assignment
25	+0·64	0·57	0	$Fe_2^{III}(C_2O_4)_3.6H_2O$
200	+1·42	2·27	0	$Fe^{II}(C_2O_4)$
300	+0·56	—	487	Fe_2O_3
500	+0·56	—	490	Fe_2O_3

P. K. Gallagher and C. R. Kurkjian, *Inorg. Chem.*, 5, 214 (1966).

After evolution of H_2O at around 100°C, a large quadrupole splitting and centre shift is observed on heating at 200°C. These parameters are characteristic of Fe^{II} high spin, and combined with the weight loss data, the decomposition product can be assigned to $Fe(C_2O_4)$. On further heating, a six-line magnetic spectrum is observed which is characteristic of Fe_2O_3. The decomposition may be represented:

$$Fe_2^{III}(C_2O_4)_3.6H_2O \xrightarrow{\sim 100°} Fe_2^{III}(C_2O_4)_3 + 6H_2O \quad (4.1a)$$

$$Fe_2^{III}(C_2O_4)_3 \xrightarrow{\sim 200°} 2Fe^{II}(C_2O_4) + 2CO_2 \quad (4.1b)$$

$$2Fe^{II}(C_2O_4) \xrightarrow{\sim 300°} Fe_2O_3 + 2CO + CO_2 \quad (4.1c)$$

Mössbauer spectroscopy

Using mainly the centre shift as a guide to the oxidation state of iron (Table 4.1) and weight losses from thermal analyses, the complex decomposition scheme of $Sr_3[Fe^{III}(C_2O_4)_3]_2.2H_2O$ has been elucidated. Table 4.5 shows that the centre shift varies from the initial value of 0·65 mm s^{-1} (Fe^{III}), to 1·44 mm s^{-1} (Fe^{II}) after heating to 200°C, to 0·60 mm s^{-1} (Fe^{III}) at 400°C, to 0·27 mm s^{-1} (Fe^{IV}) and 0·82 mm s^{-1} (Fe^{III}) at temperatures of

Table 4.5 Room temperature Mössbauer parameters (mm s^{-1}) for the thermal decomposition of $Sr_3[Fe(C_2O_4)_3]_2.2H_2O$

Temperature of heating, °C	C.S.	Q.S.	Assignment
25	+0·65	0·44	$Sr_3[Fe^{III}(C_2O_4)_3]_2.2H_2O$
200	+1·44	2·3	$Sr_3Fe_2^{II}(C_2O_4)_5$
400	+0·60	0·70	
600	+0·44	0·74	$SrCO_3.Fe_2^{III}O_3$
700	+0·27	~0	$SrCO_3.2SrFe^{IV}O_{2.8}$ (y)
	+0·82	~0	$Sr_3Fe_2^{III}O_6$ (z)
1000	+0·23	~0	(y)
	+0·69	~0	(z)

P. K. Gallagher and C. R. Kurkjian, *Inorg. Chem.*, **5**, 214 (1966).

700°C and above. A mechanism which is consistent with these oxidation states of iron and the weight loss data is:

$$Sr_3[Fe^{III}(C_2O_4)_3]_2.2H_2O \xrightarrow{100°-175°} Sr_3[Fe^{III}(C_2O_4)_3]_2 + 2H_2O \quad (4.2a)$$

$$Sr_3[Fe^{III}(C_2O_4)_3]_2 \xrightarrow{175°-350°} Sr_3[Fe_2^{II}(C_2O_4)_5] + 2CO_2 \quad (4.2b)$$

$$Sr_3[Fe_2^{II}(C_2O_4)_5] \xrightarrow{300°-450°} 3SrCO_3.Fe_2O_3 + 5CO + 2CO_2 \quad (4.2c)$$

$$3SrCO_3.Fe_2O_3 \xrightarrow{500°-650°} SrCO_3.2SrFe^{IV}O_{2.8} + 0.6CO + 1.4CO_2 \quad (4.2d)$$

$$SrCO_3.2SrFe^{IV}O_{2.8} \xrightarrow{700°-1000°} Sr_3Fe_2^{III}O_6 + \cdots \quad (4.2e)$$

The mechanism of radiolytic decomposition can be followed in a similar way. $K_3Fe^{III}(C_2O_4)_3.3H_2O$, like the strontium ferrioxalate considered in the previous example, gives one broad line with a centre shift of 0·59 mm s^{-1}. If this compound is irradiated in a 2000 Ci^{60}Co unit to a total dose of about 10^9 rads, mainly two compounds are produced. Both have the large centre shift and quadrupole splitting characteristic of Fe^{II} high spin compounds. Compound C is formed exclusively if the sample is not exposed to air during the irradiation, while compound D is the predominant product if the sample is exposed to air during the irradiation. If compound C is heated at 120° in air, it is converted rapidly into D. As discussed in chapter 2, the very large quadrupole splitting of D (Table 4.6) suggests that the Fe^{II} in D is in an environment which is very close to cubic symmetry.

On heating compound C or D at 400°, the characteristic six-line pattern of Fe_2O_3 is obtained. The infrared analysis and weight losses support the

Table 4.6 Room temperature Mössbauer parameters (mm s^{-1}) for the radiolytic and subsequent thermal decomposition of $K_3Fe^{III}(C_2O_4)_3 \cdot 3H_2O$

Conditions	C.S.	Q.S.	Assignment
1. $K_3Fe(C_2O_4)_3 \cdot 3H_2O$	+0.59	not resolved	
2. Irradiate 10^9 rad			
(a) Air excluded (C)	+1.48	2.55	$K_2Fe^{II}(C_2O_4)_2(H_2O)_2$
(b) Open to air (D)	+1.45	3.84	$K_6Fe_2^{II}(C_2O_4)_5$
3. Heat 2(a) or 2(b) to 120°C in air	+1.45	3.84	$K_6Fe_2^{II}(C_2O_4)_5$
4. Heat to 400°	+0.61	small + magnetic hyperfine splitting	Fe_2O_3

G. M. Bancroft, K. G. Dharmawardhena and A. G. Maddock, *J. Chem. Soc.* (A), 2914 (1969).

assignments given in Table 4.6, and the stoichiometry of the process can be written:

$$2K_3Fe^{III}(C_2O_4)_3 \cdot 3H_2O \xrightarrow{Vacuo} 2K_2Fe^{II}(C_2O_4)_2(H_2O)_2(C) + K_2C_2O_4$$
$$+ 2H_2O + 2CO_2 \quad (4.3a)$$

$$2K_2Fe^{II}(C_2O_4)_2(H_2O)_2 + K_2C_2O_4 \xrightarrow{120° \text{ air}} K_6[(C_2O_4)_2Fe^{II}(O_2C_2O_2)Fe^{II}$$
$$(C_2O_4)_2](D) + 4H_2O \quad (4.3b)$$

$$2K_2Fe^{II}(C_2O_4)_2(H_2O)_2 + K_2C_2O_4 + 3O_2 \xrightarrow{400°} 3K_2CO_3 + Fe_2O_3$$
$$+ 4H_2O + 7CO_2 \quad (4.3c)$$

Of particular interest in this study is the binuclear compound D which contains an unusual quadridentate oxalate group. The structure of this compound is consistent with the infrared data and the large quadrupole splitting, which as stated above, suggests that the ligand symmetry about the iron atom is close to octahedral.

These examples indicate the use of Mössbauer in elucidating solid state decompositions. However, as indicated in these examples, other data such as weight losses, infrared spectra and analyses are always needed for the complete interpretation of the decompositions.

Mössbauer spectroscopy

4.4 The effect of temperature and pressure on the electronic structure of iron compounds

A number of very interesting changes in the electronic state of the iron atom in Fe^{II} and Fe^{III} compounds have been observed by Mössbauer spectra at low temperatures or high pressures. These changes are normally reversible, and lead to a more detailed understanding of the electronic structure of these compounds.

Of particular interest has been the study of high spin-low spin equilibria in Fe^{II} compounds. A high spin compound will be preferred if the crystal field splitting Δ_0 (Fig. 2.6) is smaller than the energy required to pair two electrons. If a compound contains ligands low in the spectrochemical series (such as Cl^-), the compound is usually high spin; if it contains ligands such as CN^- which are high in the spectrochemical series, the compound will be low spin. For example, in the series of Fe^{II} bis-phenanthroline complexes $Fe(phen)_2X_2$, if $X = Cl^-$, I^-, N_3^- or OCN^-, high spin compounds are obtained having, first, the large centre shift and quadrupole splitting (Table 4.7) characteristic of Fe^{II} high spin compounds, and second, magnetic moments ($\mu = 5.0$–5.3 B.M.) slightly higher than those expected for four unpaired electrons. In contrast, if the ligand X is CN^-, NO_2^- or CNO^-, low spin compounds are obtained with, first, the small centre shift and quadrupole splitting characteristic of Fe^{II} low spin compounds and second,

Table 4.7 Mössbauer parameters for $Fe(phen)_2X_2$ and related compounds (in mm s^{-1})

Compound	T, K	C.S.	Q.S.	Type of Fe^{II}
$Fe(phen)_2(NO_2)_2$	293	0.53	0.38	low spin
$Fe(phen)_2Cl_2$	293	1.21	3.00	high spin
$Fe(phen)_2(NCS)_2$	293	1.23	2.67	high spin
	77	0.62	0.34	low spin
$Fe(phen)_2(NCSe)_2$	293	1.28	2.52	high spin
	77	0.60	0.18	low spin
$Fe(phen)_2mal.7H_2O$	293	0.59	0.18	intermediate spin
	77	0.52	0.18	
$Fe(phen)_2F_2.4H_2O$	293	0.58	0.21	intermediate spin
	77	0.55	0.16	
$Fe(pyim)_3(ClO_4)_2(E)$	294	0.58	0.53	low spin
	80	0.66	0.60	
$Fe(pyim)_3(ClO_4)_2(F)$	294	1.28	2.29	high spin
	80	0.71	0.75	low spin
		1.36	2.69	high spin

E. Konig and K. Madeja, *Inorg. Chem.*, **6**, 48 (1967).
E. Konig and K. Madeja, *Inorg. Chem.*, **7**, 1848 (1968).
D. M. L. Goodgame and A. A. S:C. Machado, *Chem. Comm.*, 1420 (1969).

the very small magnetic moments expected for no unpaired electrons. One might expect that if X is a ligand intermediate in the spectrochemical series, a compound or compounds could be prepared in which the high spin and low spin configurations are of very similar energy and in the thermally accessible range, or that an intermediate spin compound (two unpaired electrons) could be formed.

Konig and Madeja showed that for $Fe(phen)_2(NCS)_2$ and $Fe(phen)_2$-$(NCSe)_2$, the large centre shift and quadrupole splitting (Table 4.7) at room temperature are characteristic of high spin Fe^{II}, while at liquid N_2 temperatures the small centre shift and quadrupole splitting are characteristic of Fe^{II} low spin. The magnetic moment dropped from ~ 5 B.M. to 0·1 B.M. as expected, and a detailed study of the magnetic susceptibility with temperature showed that the transition point for the high spin–low spin inversion is 174 K and 223 K for the NCS and NCSe compounds respectively. This same type of equilibrium has been observed using Mössbauer spectra of a variety of other compounds. Most of these studies indicate that there is a change in molecular dimensions at the transition point.

For the compounds $Fe(phen)_2mal.7H_2O$ and $Fe(phen)_2F_2.4H_2O$, similar parameters to those of the low spin compound $Fe(phen)_2(NO_2)_2$ are obtained at both liquid N_2 and room temperatures (Table 4.7). However, other evidence from magnetic susceptibilities and electronic spectra suggests that these two compounds have an intermediate spin Fe^{II} configuration with two unpaired electrons. This result indicates once again the importance of using other techniques in conjunction with Mössbauer.

Another interesting high spin–low spin example is given by the compound $Fe(pyim)_3(ClO_4)_2$. Goodgame showed from Mössbauer and magnetic measurements that this compound has two magnetic isomers (Fig. 4.9). One compound (E) (dark blue) gives a centre shift and quadrupole splitting both at room and liquid N_2 temperatures (Table 4.7), which are characteristic of Fe^{II} low spin, and the very small magnetic moment characteristic of

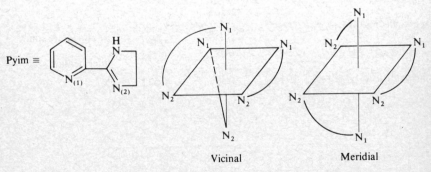

Fig. 4.9 Vicinal and meridial isomers of $Fe(pyim)_3 (ClO_4)_2$.

Mössbauer spectroscopy

$S = 0$ states. The other compound (F), which has a different X-ray powder photograph than (E) and is dark purple, has the large centre shift and quadrupole splitting, and a magnetic moment (5·25 B.M.) at room temperature characteristic of Fe^{II} high spin. At 80 K, the magnetic moment decreases to 2·7 B.M., and the Mössbauer spectrum shows two sets of absorptions, one due to Fe^{II} high spin, and the other due to Fe^{II} low spin (Table 4.7). These two compounds are probably geometric isomers. For this compound, there are two possible geometric isomers, the all *cis* (vicinal) isomer and the *trans* (meridial) isomer (Fig. 4.9). These two isomers should give small differences in the ligand field strengths at the iron atom, although it is difficult to assign the isomers.

Drickamer and co-workers have done a great deal of work on the effect of high pressure on a wide variety of iron-containing chemicals, and these studies again indicate the value of the Mössbauer effect for detecting different types of iron atoms. Perhaps the most interesting observation of these high-pressure studies is the reduction of ferric iron to ferrous with pressure. This phenomenon is reversible, with some hysteresis. Typical spectra of $Fe^{III}(AcAc)_3$ (AcAc = acetylacetone) at a range of pressures are shown in Fig. 4.10. The initial spectrum of $Fe(AcAc)_3$ is very similar to that of

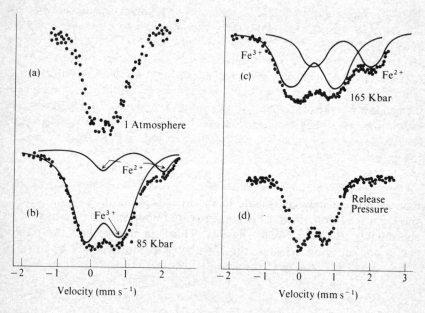

Fig. 4.10 Mössbauer spectra of $Fe(AcAc)_3$. (a) At 1 atmosphere, (b) at 85 kbar, (c) at 165 kbar, and (d) at 1 atmosphere. (From: A. R. Champion, R. W. Vaughan, and H. G. Drickamer, *J. Chem. Phys.*, **47**, 2583 (1967).)

$K_3Fe^{III}(C_2O_4)_3$—one broad peak having a centre shift of about 0·55 mm s^{-1}. On application of 85 kbar pressure, a shoulder is observed on the high velocity side of these spectra which increases in intensity at 165 kbar. To obtain a reasonable centre shift value ($<1\cdot5$ mm s^{-1}) for the new species, there must be another peak strongly overlapping the initial Fe^{III} peak. The large centre shift (>1 mm s^{-1}) gives good evidence that this new species is Fe^{II}. On releasing the pressure, the Fe^{II} peaks disappear, but the line widths of the Fe(AcAc)$_3$ peaks narrow somewhat, giving a resolvable quadrupole splitting and a similar centre shift to that observed before pressure was applied. Successive spectra at the same pressure are reproducible, while the spectrum changes rapidly with a change in pressure. This indicates that we are looking at an equilibrium phenomenon.

The above results have been interpreted in terms of a charge transfer from ligand to metal. A ferrous ion is created in a ferric site, along with a radical, radical ion, or possibly an excitation spread over several ligands. There must be a charge redistribution, as well as local compression and distortion which affect neighbouring ions, yet the complex does not break down. From a thermodynamic analysis, heats of reaction for such reactions have been obtained.

4.5 Frozen solution studies

The Mössbauer effect cannot be observed in a liquid because the f factor falls to zero (chapter 1). However spectra have been observed in glasses and smectic liquid crystals. Thus it is possible to obtain information on the structure of a molecule in solution, by taking a spectrum of a glass at, for example, 80 K. There are, however, several difficulties with this method which should be emphasized. First, one is never absolutely sure whether a true glass has been obtained, or whether the solid has crystallized. Microcrystalline domains could be formed which are exceedingly difficult to detect. It is normally best to cool the solution extremely rapidly. One technique involves spraying the solution directly onto the cold window. It is often useful to examine the sample under a microscope, although if the crystals are small enough they still may not be seen. Second, it is sometimes possible that the solvent may coordinate to the Mössbauer atom, and the desired spectrum is not obtained. This point is very important for Sn which often coordinates to rather 'weak' ligands. Third, in interpreting the results, one is not sure how sensitive the Mössbauer parameters will be to small differences in structure of the compound between solid and solution.

A few examples illustrate the usefulness and difficulties of frozen solution spectra. Goldanskii and co-workers studied compounds of the type R$_3$SnR' where R' is 4-thiopyridone or thiophenol and R is C_2H_5 or C_6H_5 (Table 4.8). The thiopyridone derivatives give quadrupole splittings in the solid state at least 1 mm s^{-1} larger than the thiophenol derivatives. These very large

Mössbauer spectroscopy

differences would not be expected if both types of compounds were four coordinate, even if the nitrogen is coordinating in the thiopyridone compounds. It was postulated that the thiopyridone compound was five coordinate with both the S and N coordinating to different Sn atoms, while the thiophenol compound is four coordinate with just the sulfur coordinating to the Sn atom. The solution spectra in pyridine of these compounds confirmed this interpretation. As seen in Table 4.8, the thiopyridone quadrupole

Table 4.8 Mössbauer parameters for R_3SnR' compounds and their pyridine solutions—All at 78 K (mm s^{-1})

R	R'	Solid		Pyridine Solution	
		C.S.	Q.S.	C.S.	Q.S.
C_2H_5	(thiocyclohexyl)	1.62	2.07	1.65	2.85
C_6H_5		1.40	1.16	1.52	2.34
C_2H_5	(thiopyridone)	1.55	3.05	1.62	3.08
C_6H_5		1.37	2.61	1.36	2.54

A. N. Nesmeyanov, V. I. Goldanskii, U. V. Khrapov, V. Ya. Rochev, D. N. Kravtsov and El. M. Rokhlena, *Bull. Acad. Science USSR*, **4**, 763 (1968).

splitting in solution is very similar to those in the solid, while the thiophenol quadrupole splitting increased by about 1 mm s^{-1}. Thus the coordination to the Sn in solution remains the same as in the solid in the thiopyridone derivatives; while the pyridine probably coordinates to the Sn in the thiophenol compounds to make them five coordinate in solution. The very similar quadrupole splitting in solution for both the thiophenol and thiopyridone compounds reflects the same coordinating moieties in both compounds in solution—three methyls, one pyridine nitrogen and a sulfur.

When unusual bond angles or lengths are observed in a compound, it is questionable whether these unusual structural parameters are due to intermolecular stacking processes, or whether they arise from the specific nature of the chemical bonding. In the compound $[\pi\text{-}C_5H_5Fe(CO)_2]SnCl_3$, the Fe—Sn bond length of 2·477 Å appeared to be shorter than any previously reported tin-transition metal bond length. For example, in the compound $[(CH_3)_4Sn_3Fe(CO)_4]_4$, the two Sn–Fe bond lengths are 2·75 Å and 2·63 Å respectively. Most other Sn–Fe bond lengths lie between 2·53 and 2·65 Å. Both Sn and Fe Mössbauer spectra for this compound are essentially the same in the solid and in frozen solution (Table 4.9). This result indicates, but does not prove, that the unusual Sn–Fe bond length is due to intramolecular bonding rather than intermolecular stacking forces. If stacking

Table 4.9 Mössbauer parameters at 80 K for
$[(\pi\text{-}C_5H_5)Fe(CO)_2]_2SnCl_2$ and $(\pi\text{-}C_5H_5)Fe(CO)_2SnCl_3$
(in mm s^{-1})

Absorber	Matrix	Resonance	C.S.	Q.S.
$[(\pi\text{—}C_5H_5)Fe(CO)_2]_2SnCl_2$	Neat	Fe	0·36	1·68
		Sn	1·95	2·38
	P*	Fe	0·35	1·74
	P	Sn	1.96	2.25
			1·31	2·29
$(\pi\text{—}C_5H_5)Fe(CO)_2SnCl_3$	Neat	Fe	0·40	1·86
		Sn	1·74	1·77
	P	Fe	0·38	1·78
	P	Sn	1·66	1·78

* P = polymethylmethacrylate solution. R. H. Herber and Y. Goscinny, *Inorg. Chem.*, **7**, 1293 (1968).

forces were important, one might expect a marked change in the Sn–Fe bond length in solution and significantly different Mössbauer spectra in the solid and solution.

The compound $[\pi\text{-}C_5H_5Fe(CO)_2]_2SnCl_2$ also gives a very short Fe–Sn bond length (2·492 Å) and long Sn–Cl bond lengths. In this compound, the SnCl$_2$ group bridges two Fe(CO)$_2(\pi\text{-}C_5H_5)$ moieties. The iron Mössbauer spectra are identical within the error in solid and solution (Table 4.9), but the tin solution spectrum splits into two sets of lines—one having very similar Mössbauer parameters to that of the solid, and one having substantially different parameters. These two sets of lines have been assigned to two conformational isomers, although the very large difference in centre shift (0·65 mm s^{-1}) for the two isomers is very surprising. It is possible that the new set of lines in the solution spectrum is the actual solution species, or more likely that the solvent has coordinated to the Sn atom.

Another example concerns the structure of Fe$_3$(CO)$_{12}$ in solution. Infrared evidence indicates that the structure is different in solution than in the solid, but attempts to obtain a different Mössbauer spectrum from that of the solid (Fig. 4.3) have not met with success. The insolubility of Fe$_3$(CO)$_{12}$ in most inert solvents makes it difficult to prepare a good absorber, and considerable difficulties have been encountered in trying to prepare a true glass.

4.6 Emission spectra

The great majority of Mössbauer spectra have been taken using a standard constant single-line source, with the absorber as the compound under study. However, a number of interesting papers have appeared in which a standard single-line absorber has been used, with the source compound being of interest. Because the source is radioactive, extensive rearrangement of the electrons and bond breaking may occur on the Mössbauer time scale (the half life of the excited state). As a result, the observed Mössbauer spectrum may not reflect the initial electronic and/or ligand environment about the Mössbauer atom. The differences in spectra can be very useful to the radiochemist in investigating the effects of the radioactive process (ref. 4.3); but for the inorganic chemist, it is desirable that the spectrum does reflect the initial electronic and ligand environment. If so, emission spectra can be very useful to the inorganic chemist. Some compounds that can be prepared using the source atom cannot be prepared using the absorber atom, and it is then possible to study the electronic environment of the artificially produced compound. For example, the iron analogue of vitamin B_{12} is not known, along with a number of more simple cobalt compounds. $XeCl_4$ (from $KICl_4$) was first 'prepared' and studied by Mössbauer. It has also been noted that in emission spectra of large molecules, a much smaller number of ^{57}Co atoms are required than ^{57}Fe in the equivalent absorber. However, there are often severe disadvantages with emission spectroscopy. Besides the difficulty with fragmentation mentioned earlier, all compounds must be prepared with small amounts of expensive radioactive isotopes. Doping experiments have often not proved to be very successful.

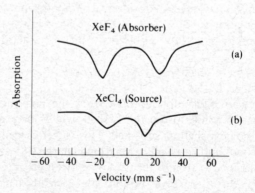

Fig. 4.11 Mössbauer spectra of: (a) absorber XeF_4 ($K^{129}IO_4$ as source) and (b) source $XeCl_4$ using $K^{129}ICl_4 \cdot H_2O$ as source and xenon clathrate as absorber. (After: G. J. Perlow and M. R. Perlow, *J. Chem. Phys.* **41**, 1157 (1964).)

Mössbauer spectroscopy as a fingerprint technique in inorganic chemistry

The first and most noteworthy contribution in this area comes from the field of xenon chemistry. The excited Xe level is populated by β decay of ^{129}I, and fortunately the majority of the decays result in no change in the number of electrons on the emitting atom. In the ^{129}I → ^{129}Xe decay in an iodine compound, the electronic structure and geometry should then be appropriate for the formation of a xenon compound having the same structure—if it is stable for a time comparable to the life time of the excited nuclear state ($\sim 10^{-9}$ sec).

A number of absorber spectra were taken with $K^{129}IO_4$ as a source which gives a single line XeO_4 on decay. Using this source and XeF_4 as absorber, a large quadrupole splitting (Fig. 4.11a) was observed, characteristic of the large p orbital imbalance one would expect from a square planar compound. Then, using $K^{129}ICl_4$ as source, and a single line Xe clathrate as absorber, the spectrum shown in Figure 4.11b was obtained. This can be attributed to the $XeCl_4$ formed on decay of the source. It only has to be stable for a time greater than 10^{-9} secs to be detected in this way. The larger quadrupole splitting in XeF_4 can be attributed to the greater p electron imbalance due to the greater electronegativity of F. Thus $N_{p_z} - \frac{1}{2}(N_{p_x} + N_{p_y})$ is larger in XeF_4 than $XeCl_4$ (chapter 6.1).

Some source work using ^{193}Os compounds as sources and a single line ^{193}Ir absorber has also provided interesting results (ref. 4.4). The 73 keV excited nuclear energy level of ^{193}Ir is populated in the β decay of ^{193}Os. As in the above ^{129}I → ^{129}Xe case, the majority of the Ir atoms should end up in a valence state which is higher by unity than the parent Os atom. However, about ten per cent of the 73 keV γ rays in ^{193}Ir are internally converted (chapter 1), leaving the Ir in a highly ionized state. The source spectra of OsO_4, $K_2OsO_4.2H_2O$, and $Os(C_5H_5)_2$ appear to be characteristic of IrO_4^+, IrO_4^-, and $Ir(C_5H_5)_2^+$ respectively, and provide valuable information on unknown and highly unstable iridium compounds. Unfortunately, in a number of other compounds such as K_2OsX_6, the expected Ir^V species is not observed. A complex spectrum is obtained which is rather difficult to interpret.

In other forms of radioactive decay, more substantial electronic rearrangement takes place. For example, ^{57}Co captures an orbital electron to give ^{57}Fe. This electron is usually captured from the innermost shell, the K shell. This vacancy is filled from the L shell, which in turn is filled by an electron from the next shell or sub-shell. This Auger cascade takes place in about 10^{-4} seconds and can lead to very high charge states on the resulting iron atom. When the Co atom is incorporated in a molecule, the high charge states may be neutralized very rapidly by electrons from other atoms, which may lead to atomic rearrangements and bond breaking. If ^{57}Co is placed in a metallic environment such as stainless steel or palladium, the above effects vanish in a time which is short compared with the $t_{1/2}$ of

$\sim 10^{-7}$ secs. This results in a single narrow Mössbauer line which coincides in energy with the absorption line of ^{57}Fe in stainless steel or palladium respectively. Thus using a ^{57}Co in stainless steel source, and a stainless steel absorber, a single fairly narrow line at zero relative velocity is observed. In some molecular compounds, the Auger process causes degradation of the molecule and a change in the valence state of the iron atom. $Co^{III}(d^6)$ should yield $Fe^{III}(d^5)$ if the Auger process does not disturb the molecular structure. In $^{57}Co(AcAc)_3$, appreciable Fe^{II} is formed while the main absorption is characteristic of $Fe^{III}(AcAc)_3$. In other molecules, the absorptions are not characteristic of expected Fe^{II} or Fe^{III} species and very broad lines are observed. Yet in molecules such as vitamin B_{12} and Co^{II}-phthalocyanine, [$Co^{II}Pc$], the spectra are characteristic of Fe^{II} with very little, if any effect from fragmentation. For example, the $Co^{II}Pc$ spectrum agrees reasonably well with the absorber spectrum of $Fe^{II}Pc$. (ref. 4.5).

It is apparent from the above work that with many Mössbauer sources, a good deal of background work is required to ascertain whether most molecules do fragment during the radioactive decay processes. For β decay, as in $^{129}I \rightarrow Xe$ and $^{193}Os \rightarrow Ir$, useful results to the inorganic chemist have, and will be, obtained. For electron capture, such as in $^{57}Co \rightarrow Fe$, it appears that the majority of compounds will fragment and/or give charge states which are not characteristic of the expected species.

Problems

1. Two analogous compounds $(\pi\text{-}C_5H_5)CoFe_2(CO)_9$ and $(\pi\text{-}C_5H_5)RhFe_2(CO)_9$ have recently been prepared [J. Knight and M. J. Mays, *J. Chem. Soc.*, A, 654, (1970)]. However, the first compound gives an ^{57}Fe Mössbauer spectrum consisting of four lines of equal intensity, while the latter compound gives just two narrow peaks. Suggest structures which are consistent with the above spectra and the 18 electron rule.
2. Although the structure of $Fe_3(CO)_{12}$ in the solid state is now known to be that in Figure 4.3e, the solution infrared spectra suggest that $Fe_3(CO)_{12}$ in solution has the D_{3h} structure with three single carbonyl bridges and three terminal carbonyls on each iron atom. What differences in the Mössbauer spectra between solid and solution might be expected? (C. H. Wei and L. F. Dahl, *J. Am. Chem. Soc.*, **91**, 1351 (1969); M. Poliakoff and J. J. Turner, *J. Chem. Soc.* (A), 654 (1971).
3. The compound $Fe(py)_4(NCS)_2$ exists in two forms: a yellow form and a violet form. It was concluded in 1964 that the yellow form was the *cis*-compound and the violet form, the *trans*-compound. The Mössbauer spectra of these forms gave identical spectra with quadrupole splitting

= 1·53 mm s^{-1} and centre shift = +1·42 mm s^{-1}. Give possible explanations for this observation. What electronic state is the Fe^{II} in?

4. It has long been held that Prussian blue, KFe^{III}[Fe^{II}(CN)$_6$] (made by adding Fe^{III} to Fe^{II}(CN)$_6^{4-}$), and Turnbull's blue (made by adding Fe^{II} to Fe^{III}(CN)$_6^{3-}$) were distinct compounds. The Mössbauer spectra of both compounds give three peaks: a singlet with centre shift $\sim$0·3 mm s^{-1}; and a doublet having a centre shift of $\sim$0·8 mm s^{-1}, and a quadrupole splitting of $\sim$0·4 mm s^{-1}. How would you assign the two sets of lines? Are the two compounds distinct compounds? (M. B. Robin and P. Day, *Adv. Inorg. Radiochem.*, **10**, 247 (1967); A. K. Bonnette and J. F. Allan, *Inorg. Chem.*, **10**, 1613 (1971).)

5. Addition of the bidentate ligand dppe(Ph$_2$PCH$_2$CH$_2$PPh$_2$) to FeCl$_2$ in benzene gives a white crystalline solid. This gives an ^{57}Fe Mössbauer spectrum with a centre shift of $\sim$0·86 mm s^{-1} and a quadrupole splitting of $\sim$2·75 mm s^{-1}. Is this compound low spin Fe^{II}Cl$_2$(dppe)$_2$? What are more likely possibilities? What other techniques would you use to distinguish these possibilities?

References

4.1 D. E. Fenton and J. J. Zuckerman, *Inorg. Chem.*, **8**, 1771 (1969).
4.2 G. M. Bancroft, and K. D. Butler, *J. Chem. Soc.* (A), 1209 (1972).
4.3 A. G. Maddock, in MTP Publications (1972), vol. 8. Butterworths and University Park Press.
4.4 R. P. Rother, F. Wagner, and U. Zahn, *Radiochimica Acta.*, **11**, 203 (1969).
4.5 A. Nath, M. P. Klein, W. Kundig and D. Lichtenstein, *Mössbauer Effect Methodology*, ed. I. J. Gruverman, Plenum Press, vol. 5.

Bibliography

R. H. Herber, *Prog. Inorg. Chem.*, **8**, 1 (1966).
R. V. Parish, *Prog. Inorg. Chem.*, **15**, 101 (1972).
J. J. Zuckerman, *Adv. Organomet. Chem.*, **9**, 21 (1970).
G. M. Bancroft and R. H. Platt, *Adv. Inorg. Radiochem.*, **15**, 59 (1972).

5. Centre shifts and bonding properties

In chapter 2, we discussed the dependence of the isomer shift on $\delta R/R$ and the s-electron density at the nucleus, $[\Psi(0)_s]^2$ (eq. 2.4). If $\delta R/R$ is positive, then an *increase* in the s-electron density at the absorber nucleus *increases* the isomer shift and the centre shift; if $\delta R/R$ is negative, then an *increase* in the s-electron density at the absorber nucleus *decreases* the centre shift. The s-electron density at the nucleus $[\Psi(0)_s]^2$ increases with increase in valence shell s-electron density (e.g., $3s$ or $4s$ electron density), but $[\Psi(0)_s]^2$ decreases as the p- and/or d-electron density increases. Thus for positive $\delta R/R$, an *increase* in p- and/or d-electron density will *decrease* the centre shift.

Although rigorous theoretical calculations of $[\Psi(0)_s]^2$ have been carried out for some Mössbauer isotopes (notably ^{57}Fe and ^{119}Sn) using the Fermi–Segré formula (eq. 2.5) and/or Hartree–Fock calculations, such calculations have not yet been performed for many isotopes. For the heavier isotopes such as ^{197}Au, such calculations are not yet feasible. In this chapter, we attempt to rationalize trends in centre shift values for series of similar compounds using rather simplified, but useful, molecular orbital arguments and/or to correlate the centre shift values with other more well established empirical parameters such as electronegativity, the spectrochemical series or the nephelauxetic series. The above correlations lead to a consistent (if oversimplified) picture of trends in centre shift values for a large number of Mössbauer ions. In addition, valuable predictions on the bonding properties of such novel ligands as N_2 and the electronegativities of such ligands as N_3^- can be made.

We begin this chapter by discussing centre shift values for simple iodine and xenon compounds. For both of these $\delta R/R$ is positive, and we would expect that the centre shift would increase with increase in $5s$-electron density, but decrease with an increase in the $5p$- and/or $4d$-electron densities. Unlike the other Mössbauer isotopes discussed in this chapter, the centre shift values are sensitive to the p-electron density changes (rather than changes in s-electron density), presumably because the bonding in these

Centre shifts and bonding properties

compounds involves mainly Xe or I p orbitals rather than 'spd hybrids'. In contrast, several simple series of Fe^{II} low spin and Ru^{II} compounds show that the centre shift is sensitive to changes in 4s- and 5s-electron densities respectively due to the σ bonding ligands. Thus a strong σ donor ligand such as CN^- or H^- increases the 4s- or 5s-electron densities, leading to an increase in the s-electron density at the nucleus $[\Psi(0)_s]^2$. Since $\delta R/R$ is negative for iron but positive for Ru, the above σ donor ligands will give a comparatively negative centre shift for Fe, but a comparatively positive centre shift for Ru. Then we turn to Sn ($\delta R/R = +$ve), and Sb($\delta R/R = -$ve) and correlate centre shift values with electronegativities and bonding properties. Finally, the concept of partial centre shifts (p.c.s.) is critically examined using the data for Sn^{IV}, Fe^{II} low spin, Fe^{2+} high spin and Au ($\delta R/R = +$ve) compounds to illustrate the difficulties and potential of the partial centre shift concept. Useful correlations between the partial centre shift values and the spectrochemical and nephelauxetic series are discussed.

5.1 Correlation of centre shifts with electronegativities and σ and π bonding properties of ligands

Iodine and Xenon Compounds. Centre shifts for nine simple iodine compounds are given in Table 5.1. The first four compounds give positive centre shift values, while the latter five (the alkali halides) give negative values. Since $\delta R/R$ is positive for ^{129}I, we can immediately say that $[\Psi(0)_s]^2$ is larger for the first four compounds than the alkali halides. This implies that the 5s orbital population is larger and/or that the 5p orbital population is smaller for the first four compounds.

The bonding in iodine compounds can be explained formally in two contrasting ways: participation of just the iodine 5p orbitals in bonding, or participation of iodine 'spd hybrids' in bonding. The bond angles in many iodine compounds such as the trihalide ions I_3^- and ICl_2^- are $\sim 180°$; while the interhalogen compounds such as ICl_4^- have a square planar structure. These 90° and 180° bond angles have been taken as evidence (although not conclusive) that iodine just uses p orbitals in bonding in at least the above compounds and the interhalogens. For other simple iodine species such as IO_4^- and IO_3^-, iodine 'spd hybrid' orbitals are probably used in the bonding to oxygen.

The centre shift values in Table 5.1 are consistent with mainly iodine 5p participation in bonding. For the covalent I_2 molecule, we could classify the iodine atoms as having a 5p^5 electronic configuration, and the centre shift value for I_2 would thus correspond to the value for one p hole (h_p) where:

$$h_p = 6 - (N_{p_x} + N_{p_y} + N_{p_z}) \qquad (5.1)$$

Mössbauer spectroscopy

In contrast, the iodine in the alkali halides would be expected to be close to I^- with the electronic configuration $5p^6$. The very small values for h_p (Table 5.1) calculated from the average of previous n.m.r. chemical shift and dynamic quadrupole data are consistent with the iodine being close to I^-. Using the h_p values for compounds 1, 5, 6, 7, 8, and 9, and assuming a linear dependence of centre shift with h_p (Fig. 5.1), the following equation is obtained:

$$\text{C.S.} = 0{\cdot}136\, h_p - 0{\cdot}054\, (\text{cm s}^{-1}). \tag{5.2}$$

For a closed p shell $5p^6$, ($h_p = 0$), the centre shift equals $-0{\cdot}054$ cm s^{-1}, in agreement with earlier calculated values.

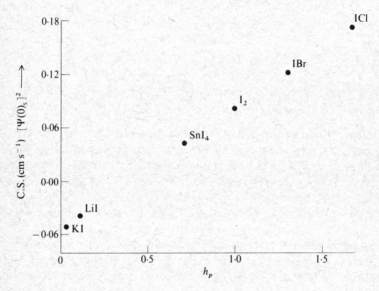

Fig. 5.1 Plot of centre shift versus h_p for iodine compounds in Table 5.1. (Based on: M. Pasternak, A. Simopoulos, and Y. Hazony, *Phys. Rev.*, **140A**, 1892 (1965).)

Using eq. 5.1, the h_p values for compounds 2, 3, and 4 are calculated (Table 5.1, Fig. 5.1). The value of h_p increases in the order of increasing electronegativity of the atom bonded to the iodine, i.e. $SnI_4 < I_2 < IBr < ICl$. Chlorine being the most electronegative, would be expected to decrease the $5p$ orbital population on the iodine (and increase h_p) to the greatest extent in these four compounds. An increase in h_p will cause a deshielding of the $5s$ electrons on I, and cause $[\Psi(0)_s]^2$ to increase. Thus ICl has the most positive centre shift.

The order of centre shift values is consistent with the assumption that the $5s$ orbitals are not utilized to any great extent in the bonding in these compounds. Consider for a moment that the iodine uses sp hybrid orbitals for

Centre shifts and bonding properties

Table 5.1 Centre shifts (relative to standard ZnTe source) for iodine compounds (cm s^{-1} at ~80 K)

Compound	C.S.	h_p
I_2	+0.083	1
IBr	+0.123	1.30*
ICl	+0.173	1.67*
SnI_4	+0.043	0.71*
LiI	−0.038	0.11
NaI	−0.046	0.06
KI	−0.051	0.03
RbI	−0.043	0.06
CsI	−0.037	0.16

* Calculated from: C.S. = $0.136 h_p - 0.054$ (cm s^{-1}).
M. Pasternak and T. Sonnino, *J. Chem. Phys.*, **48**, 1997 (1968); D. W. Hafemeister, G. DePasquali, and H. deWaard, *Phys. Rev.*, **135**, B1089 (1964); M. Pasternak, A. Simopoulos, and Y. Hazony, *Phys. Rev.*, **140**, A1892 (1965); S. Bukshpan and R. H. Herber, *J. Chem. Phys.*, **46**, 3375 (1967).

bonding in I_2, IBr, and ICl. An electronegative ligand such as Cl would thus decrease both the iodine 5s and 5p orbital populations. Since $[\Psi(0)_s]^2$ and the centre shifts are usually much more sensitive to s orbital than p orbital occupancy changes, (eq. 2.6 for Sn, and the results in the following section), we would expect that the compound with the most electronegative ligand, ICl, would have the smallest $[\Psi(0)_s]^2$ and the most negative centre shift. This effect is not observed, indicating that changes in the 5p orbital occupancy dominate the centre shifts for these iodine compounds.

In the next chapter, the values of h_p will be used in conjunction with quadrupole splitting results to derive values of N_{p_x}, N_{p_y} and N_{p_z}.

For xenon compounds (Table 5.2), the centre shift values again become generally more positive as the electronegativity of the ligands increase.

Table 5.2 Centre shifts (relative to xenon clathrate) for xenon compounds (mm s^{-1} at 4.2 K)

Substance*	C.S.	Substance*	C.S.
XeF_4	+0.40 ± 0.04	$[XeCl_2]$	+0.17 ± 0.08
$[XeCl_4]$	+0.25 ± 0.08	$[XeBr_2]$	−0.03 ± 0.07
XeF_2	+0.10 ± 0.12		

* Those compounds in brackets were 'prepared' from the corresponding I compound and used as sources. For example, $XeCl_4$ was 'prepared' from $KICl_4$ (chapter 4.6).
G. J. Perlow and H. Yoshida, *J. Chem. Phys.*, **49**, 1474 (1968).

Mössbauer spectroscopy

The errors are often comparable to the differences in centre shift values and XeF_2 has a slightly smaller centre shift than $XeCl_2$. Since $\delta R/R$ is positive again, these results are again consistent with xenon using almost pure $5p$ orbitals for bonding in these compounds.

Fe^{II}, Ru^{II} and Au^{I} compounds. In the iodine and xenon compounds discussed above, the centre shift values varied due to changes in $5p$ orbital populations rather than changes in $5s$ orbital population. Most other Mössbauer ions use a '*dsp* hybrid orbital(s)' to bond to the ligand(s), and in contrast, the centre shift is more sensitive to the changes in the s orbital occupancy, i.e. a more electronegative ligand will withdraw s-, p- and d-electron density, but $[\Psi(0)_s]^2$ and the centre shift for positive $\delta R/R$ will *decrease*.

Trends in centre shift values for Fe^{II} and Ru^{II} compounds illustrate the above ideas, and lead to useful bonding information for ligands. The molecular orbital diagram for Fe^{II} low spin (and Ru^{II} compounds) is given in Fig. 5.2.

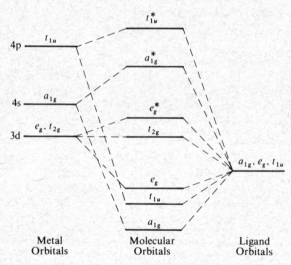

Fig. 5.2 Molecular orbital diagram for octahedral transition metal complexes. Just σ bonding is included.

Both Fe^{II} and Ru^{II} ions have a $(t_{2g})^6$ configuration. Ligand donation from appropriate σ orbitals populates the e_g, a_{1g} and t_{1u} molecular orbitals formed by overlap of the ligand orbitals with the metal d_{z^2}, $d_{x^2-y^2}$, $4s$ and $4p$ orbitals. In hybridization terms, the σ bonding can be represented by overlap between d^2sp^3 hybrids of the iron atom with the σ orbitals of the ligands. In many compounds, the ligands also have empty π^* orbitals which overlap with the metal t_{2g} electrons. Such overlap results in a donation of the t_{2g} electrons on the metal to the ligand π^* orbitals.

Both of the above mechanisms—σ donation and π acceptance—increase $[\Psi(0)_s]^2$. For σ donation, $[\Psi(0)_s]^2$ is more sensitive to 4s- or 5s-electron augmentation than d- or p-electron augmentation, and σ donation results in an increase in $[\Psi(0)_s]^2$. For π acceptance, the t_{2g} electrons become more delocalized, and shield the valence s electrons less. Once again the s-electron density at the nucleus increases. Thus strong σ donors (e.g. H$^-$) and strong π acceptors (e.g. NO$^+$) give the largest s-electron densities at the Fe^{II} and Ru^{II} nuclei. Because $\delta R/R$ is negative for ^{57}Fe, but positive for ^{99}Ru, an increase in $[\Psi(0)_s]^2$ causes the centre shift to decrease for ^{57}Fe, but increase for ^{99}Ru. Relative centre shift values for these two isotopes can then be used as a measure of the σ donor plus π acceptor capacities of ligands [denoted $(\sigma + \pi)$].

Centre shifts for [Fe(CN)$_5$X^{n-}]$^{(3+n)-}$ and *trans*-[FeHL(depe)$_2$]$^+$BPh$_4^-$ compounds are given in Table 5.3. For the former series, the NO$^+$ compound,

Table 5.3 Centre shifts for [Fe(CN)$_5$X^{n-}]$^{(3+n)-}$ and trans-[FeHL(depe)$_2$]$^+$BPh$_4^-$ * compounds (mm s^{-1} at 295 K)

X	C.S.	L	C.S.
NO$^+$	0.00	CO	+0.12
CO	+0.15	pMeO.C$_6$H$_4$.NC	+0.19
SO$_3^=$	+0.22	Me$_3$CNC	+0.21
P(C$_6$H$_5$)$_3$	+0.23	P(OMe)$_3$	+0.25
NO$_2^-$	+0.26	P(OPh)$_3$	+0.26
Sb(C$_6$H$_5$)$_3$	+0.26	N$_2$	+0.32
NH$_3$	+0.26	PhCN	+0.33
As(C$_6$H$_5$)$_3$	+0.29	MeCN	+0.35
H$_2$O	+0.31	Cl$^-$	+0.39
		I$^-$	+0.39

* depe = Et$_2$P.CH$_2$.CH$_2$.PEt$_2$.
E. Fluck, W. Kerler and W. Neuwirth, *Angew. Chem. Internat. Edn.*, **2**, 277 (1963); N. L. Costa, J. Danon, and R. M. Xavier, *J. Phys. Chem. Solids*, **23**, 1783 (1962); E. Fluck and P. Kuhn, *Z. Anorg. Chem.*, **350**, 263 (1967); N. E. Erickson, Ph.D. Thesis, University of Seattle, Washington, USA; G. M. Bancroft, M. J. Mays, B. E. Prater and F. P. Stefanini, *J. Chem. Soc. (A)*, 2146 (1970).

as expected, gives the most negative centre shift value: H$_2$O, a rather weak σ donor with little or no π acceptor ability gives the most positive value. Thus, $(\sigma + \pi)$ increases in the order H$_2$O < As(C$_6$H$_5$)$_3$ < NH$_3$ ~ Sb(C$_6$H$_5$)$_3$ ~ NO$_2^-$ < P(C$_6$H$_5$)$_3$ < SO$_3^=$ < CO $\ll$ NO$^+$. Later results (Table 5.12) indicate that the centre shift has a similar sensitivity to both σ donation and π acceptance. Thus both H$^-$ and NO$^+$ give comparable decreases in centre shift values in Fe^{II} compounds (chapter 5.2).

In the above compounds, and in other series of compounds in which just one or two ligands are changed, we assume that the bonding properties of

Mössbauer spectroscopy

the fixed ligands (CN^- in the above series) do not change appreciably and that the centre shift is independent of the cation or anion in anionic or cationic compounds respectively. Thus we assume that the changes in centre shift are due mainly to the different bonding properties of the changed ligands. These assumptions appear to be valid for the great majority of compounds.

For the *trans*-$[FeHL(depe)_2]^+ BPh_4^-$ compounds, the centre shift *decreases* in the order $I^- > Cl^- > MeCN > PhCN > N_2 > P(OPh)_3 > P(OMe)_3 > Me_3CNC > $ p-$MeO.C_6H_4.NC > CO$ (Table 5.3) indicating that $(\sigma + \pi)$ *increases* in the above order. This order is consistent with much chemical data. For example CO is known to be a good $(\sigma + \pi)$ ligand, whereas I^- and Cl^- are comparatively poor $(\sigma + \pi)$ ligands, giving bonds with more ionic character. The above series is of particular interest because of the preparation of the novel N_2 compound. The order indicates that N_2 is a comparable $(\sigma + \pi)$ ligand to the nitriles MeCN and PhCN. The ruthenium compounds discussed shortly also indicate that N_2 is a comparable ligand to the nitriles, and present preparative data indicate that N_2 compounds can usually be prepared if the corresponding nitrile compound is known. The quadrupole splitting data reported in the next chapter (Table 6.3) indicate that N_2 is a weak σ donor but a moderate π acceptor.

Table 5.4 Centre shifts (relative to Ru metal) for Ru^{II} compounds (mm s^{-1} at 4·2 K)

Compound	C.S.*	Compound	C.S.
$[Ru(NH_3)_6]Cl_2$	−0·82	$K_2[Ru(CN)_5NO]$	−0·06
$[Ru(NH_3)_5(CH_3CN)](ClO_4)_2$	−0·71	$Rb_2[Ru(NCS)_5NO]$	−0·30
$[Ru(NH_3)_5N_2]Cl_2$	−0·62	$Rb_2[RuCl_5NO]$	−0·37
$[Ru(NH_3)_5SO_2]Cl_2$	−0·61	$Cs_2(RuBr_5NO)$	−0·47
$[Ru(NH_3)_5CO]Br_2$	−0·54		
$[Ru(NH_3)_5NO]Cl_3 . H_2O$	−0·18		

* Average values from references below.
W. Potzel, F. E. Wagner, U. Zahn, R. L. Mössbauer, and J. Danon, *Z. Physik*, **240**, 306 (1970); R. A. Prados, *Ph.D. Thesis*, Louisiana State University, New Orleans (1971) (kindly communicated by M. L. Good); R. Greatrex, N. N. Greenwood, and P. Kaspi, *J. Chem. Soc. (A)*, 1873 (1971).

For the $[Ru^{II}(NH_3)_5X^{n-}]$ compounds (Table 5.4), because $\delta R/R$ is positive, the centre shift increases as $(\sigma + \pi)$ increases. Thus, $(\sigma + \pi)$ increases in the order $NH_3 < CH_3CN < N_2 < SO_2 < CO \ll NO^+$. Again this series indicates that N_2 is a comparable $(\sigma + \pi)$ ligand to MeCN, but that CO and NO^+ are much better σ donors and/or π acceptors. This series does indicate that N_2 is a better $(\sigma + \pi)$ ligand than NH_3, presumably because of the stronger π acceptor ability of N_2.

For the $[RuX_5(NO)]^=$ series, the centre shift and thus $(\sigma + \pi)$ increases in the order $X = Br^- < Cl^- < NCS^- < CN^-$, again in agreement with the known strong σ donor properties of CN^-. This order shows a strong correlation with the ranking of these ligands in the spectrochemical series. Such correlations will be discussed in the next section of this chapter.

For Au^I compounds of the type LAuCl, the centre shift is very sensitive to L (Table 5.5). Assuming that sp hybridization is important in these compounds, and that, like Fe and Ru, the centre shift is more sensitive to variations in the $6s$ electron density than the $6p$, then neglecting π bonding, the centre shift should become more positive ($\delta R/R$ is positive) as the σ

Table 5.5 Centre shifts (relative to the source Au*Pt*) for gold compounds (mm s^{-1} at 4·2 K)

Compound	C.S.	Compound	C.S.
$C_6F_5Ph_2PAuCl$	2·93	C_5H_5NAuCl	1·7
Ph_3PAuCl	2·96	Me_2SAuCl	1·26
$Ph_3AsAuCl$	1·92	AuCl	−1·2

J. S. Charlton and D. I. Nichols, *J. Chem. Soc. (A)*, 1484 (1970).

donor capacity of the ligands increase. Thus, for the compounds in Table 5.5, the σ donor ability increases in the order $Me_2S < C_5H_5N < Ph_3As < Ph_3P < (C_6F_5)Ph_2P$.

Sn^{IV}, Sb^{III} and Fe^{II} high spin compounds. In addition to the correlations of centre shift with electronegativities noted for xenon and iodine compounds, several other good correlations of centre shift with electronegativities have been noted. In this section, we will examine three of these correlations, before looking at results for Sn^{IV} compounds which are rather difficult to rationalize—even using the empirical approach outlined in this and previous sections.

The signs of $\delta R/R$ for Sn and Sb are positive and negative respectively, so that an increase in s-electron density at the nucleus gives a more positive centre shift for Sn, but a more negative centre shift for Sb.

Sn and Sb have the electronic configurations $5s^2 5p^2$ and $5s^2 5p^3$ respectively. The bonding in four-coordinate Sn^{IV} compounds can be described by two extreme models: first, donation of ligand electrons to the empty sp^3 hybrids of a 'bare' Sn^{4+} ion; or second, formation of covalent Sn^{IV} ($5s5p^3$) with covalent bonding to the ligands. In six-coordinate Sn^{IV} compounds, the bonding probably involves d^2sp^3 hybrids. For Sb^{III} (formally $5s^2$), the bonding probably again involves 'sp hybridization', as there is little evidence for a 'lone' $5s^2$ pair.

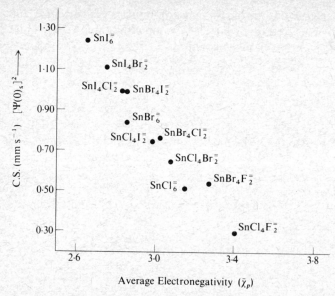

Fig. 5.3 Plot of centre shift versus the average of Pauling electronegativities for $SnX_4Y_2^=$ compounds. (R. V. Parish, *Prog. Inorganic Chemistry*, **15**, 101 (1972).) The centre shift for $SnF_6^=$ is not included in the graph.

A good correlation exists between the centre shift and the sum of Pauling ligand electronegativities (Fig. 5.3) for $SnX_4Y_2^=$ species (Table 5.6). As the electronegativity increases, the centre shift decreases—in contrast to iodine and xenon compounds in which the centre shift increased with increase in electronegativity of the ligands. This trend in Sn^{IV} centre shift values is expected considering that the bonding involves some type of '*sp* or *spd* hybridization', and that the centre shift is most sensitive to changes in *s*-electron density. Thus the more electronegative ligands effectively withdraw

Table 5.6 Centre shifts (relative to SnO_2) for $(Et_4N)_2SnX_4Y_2$ compounds (mm s^{-1} at 80 K)

Compound	C.S.	Compound	C.S.
$SnCl_4F_2^=$	0.29	$SnBr_6^=$	0.84
$SnBr_4F_2^=$	0.53	$SnBr_4I_2^=$	0.99
$SnCl_6^=$	0.51	$SnI_4Cl_2^=$	0.99
$SnCl_4Br_2^=$	0.65	$SnI_4Br_2^=$	1.11
$SnBr_4Cl_2^=$	0.76	$SnI_6^=$	1.24
$SnCl_4I_2^=$	0.74	$SnF_6^=$	−0.36

Averages from: C. A. Clausen and M. L. Good, *Inorg. Chem.*, **9**, 817 (1970); A. G. Davies, L. Smith and P. J. Smith, *J. Organmetal. Chem.*, **23**, 135 (1970).

5s-electron density from the Sn and give more negative centre shift values. A similar correlation exists for SnX_4 compounds, and a least-squares fit of these lines gives the following equations: (ref. 5.1)

$$\text{C.S.} \quad (SnX_4)(\text{mm s}^{-1}) = 4\cdot82 - 1\cdot27\bar{\chi}_p \qquad (5.3)$$

and

$$\text{C.S.} \quad (SnX_6^=)(\text{mm s}^{-1}) = 4\cdot27 - 1\cdot16\bar{\chi}_p. \qquad (5.4)$$

Omitting the points for the fluoro complexes, equation (5.4) becomes:

$$\text{C.S.} \quad (SnX_6^=) = 4\cdot96 - 1\cdot40\bar{\chi}_p. \qquad (5.5)$$

These correlations have been used to obtain Pauling electronegativities χ_p for a number of groups (Table 5.7) using eqs. (5.3) and (5.4). For example,

Table 5.7 Electronegativities (Pauling scale) from Sn centre shifts (mm s^{-1} at 80 K)

Ligand	In compound	C.S.	χ_p
N_3^-	$[(CH_3)_4N]_2Sn(N_3)_6$	0·48	3·3
$O^=$	SnO_2	0·00	3·7
CH_3	$(CH_3)_4Sn$	1·31	2·8
C_6H_5	$(C_6H_5)_4Sn$	1·22	2·8
NMe_2	$Sn(NMe_2)_4$	0·84	3·1
NEt_2	$Sn(NEt_2)_4$	0·76	3·2

R. F. Dalton and K. Jones, *Inorg. Nucl. Chem. Lett.*, **5**, 785 (1969); R. H. Herber and H. S. Cheng, *Inorg. Chem.*, **9**, 1686 (1970); R. V. Parish, *Prog. Inorg. Chem.*, **15**, 101 (1972).

for the $Sn(N_3)_6^=$ species, $0\cdot48 = 4\cdot27 - 1\cdot16\bar{\chi}_p$. Solving, we find $\chi_p(N_3^-) = 3\cdot3$. The values obtained for the Pauling electronegativities are close to the usual values quoted for the donor atoms.

Similarly for Sb^{III} compounds, the centre shift values (Table 5.8) plotted against the difference in the Pauling electronegativity between Sb and the ligand gives a reasonable correlation—with the exception of SbF_3 (Fig. 5.4).

Table 5.8 Centre shifts (relative to $^{121}SnO_2$) for Sb^{III} compounds (mm s^{-1} at 78 K)

Compound	C.S.	Compound	C.S.
SbF_3	−14·6	SbI_3	−15·9
$SbCl_3$	−13·8	Sb_2O_3	−11·3
$SbBr_3$	−13·9		

L. Bowen, J. G. Stevens and G. G. Long, *J. Chem. Phys.*, **51**, 2010 (1969).

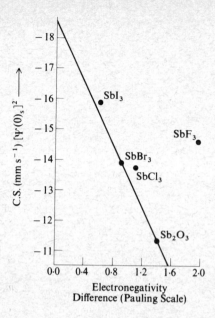

Fig. 5.4 Plot of centre shift versus electronegativity difference for some Sb^{III} compounds. (L. H. Bowen, J. G. Stevens, and G. G. Long, *J. Chem. Phys.*, **51**, 2010 (1969).)

Because $\delta R/R$ is negative for ^{121}Sb, the centre shift becomes more positive as the electronegativity of the ligands increases; but like ^{119}Sn, $[\Psi(0)_s]^2$ decreases as the electronegativity increases. This trend is again expected, assuming that the centre shift is more sensitive to 5s occupancy changes than 5p or 4d occupancy changes. It is noticeable from Figures (5.3) and (5.4), that the sensitivity of the centre shift is much larger for Sb than Sn. Thus the difference in centre shifts between Sb_2O_3 and SbI_3 is 5 mm s^{-1}, whereas the maximum range for the $SnX_4Y_2^=$ species is less than 1 mm s^{-1}. These results suggest that $\delta R/R$ is over five times as large for ^{121}Sb than ^{119}Sn (ref. 5.2).

Other such correlations of centre shifts with electronegativity have been found for aurous halides, octahedral halides of Te^{4+} and Fe^{2+} high spin halides. The latter correlation is shown in Fig. 5.5. The centre shift here has been corrected for the second order Döppler term (chapter 2.1), so that isomer shift is plotted versus electronegativity. Again we plot the isomer shift such that $[\Psi(0)_s]^2$ increases. Since $\delta R/R$ is negative for ^{57}Fe, an increase in isomer shift corresponds to a decrease in $[\Psi(0)_s]^2$. For Sn^{IV}, Sb^{III}, and Fe^{II} high spin (Figs 5.3, 5.4, and 5.5), $[\Psi(0)_s]^2$ decreases as the electronegativity of the ligands increases in contrast to the reverse trend for Xe and I (Figs 5.1 and 5.2). The order of centre shift values for iron compounds follows the electronegativity series and the nephelauxetic series, and this correlation will be discussed in the next section of this chapter.

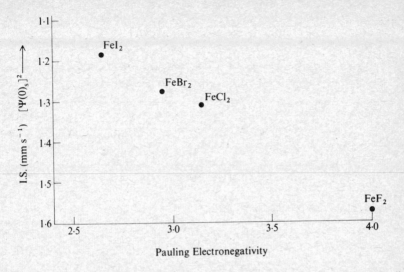

Fig. 5.5 Isomer shift (with respect to natural iron at room temperature) versus Pauling electronegativity for the ferrous halides. (R. C. Axtmann, Y. Hazony, and J. W. Hurley, *Chem. Phys. Lett.*, **2**, 673 (1968).)

'Anomalous' Sn^{IV} *centre shift results.* For many Sn^{IV} compounds—most Sn organometallics—it is not possible at first sight to draw correlations with electronegativities or σ donor capacities and centre shift data. For example, on the arguments presented in the last section, we would expect the centre shift to increase as the average ligand electronegativity decreases. However, comparing the four-coordinate Ph_3SnX (X = Cl, Br) and Ph_2SnX_2 compounds (Table 5.9), it is apparent that the centre shift is larger in Ph_2SnX_2 than Ph_3SnX. Thus the centre shift *increases* as the average electronegativity *increases*. Similarly in the $Me_{3-n}Cl_nSnMn(CO)_5$ series (Table 5.9), the centre shift again increases on substituting the electronegative Cl for the electropositive Me group. The similar trend for I and Xe compounds

Table 5.9 Centre shifts (relative to SnO_2) for organotin compounds (mm s^{-1} at 80 K)

Compound	C.S.	Compound	C.S.
Ph_3SnCl	1·34	$Me_3SnMn(CO)_5$	1·41
Ph_2SnCl_2	1·38	$Me_2ClSnMn(CO)_5$	1·52
Ph_3SnBr	1·33	$MeCl_2SnMn(CO)_5$	1·62
Ph_2SnBr_2	1·43	$Cl_3SnMn(CO)_5$	1·65

R. V. Parish and R. H. Platt, *Inorg. Chim. Acta.*, **4**, 589 (1970); G. M. Bancroft, K. D. Butler, and A. T. Rake, *J. Organomet. Chem.*, **34**, 137, (1972).

($[\Psi(0)_s]^2$ increases with increase in electronegativity) suggests one possible explanation. If the bonding in these compounds involves mostly Sn p orbitals, and/or if the s-electron density is concentrated in the Sn–Me or Sn–Mn(CO)$_5$ bonds, then the halides will withdraw mostly $5p$-electron density, thus increasing $[\Psi(0)_s]^2$ and the centre shift with successive substitution of halogens. Alternatively, it has been suggested that the electronegative halide removes charge from the valence shell, resulting in a residual positive charge on the Sn atom which produces a deshielding and contraction of the $5s$ orbitals and an increase in $[\Psi(0)_s]^2$. The former argument is perhaps more satisfying because the iodine and xenon centre shift trends can also be rationalized by assuming mostly p bonding to halides.

5.2 Partial centre shifts (p.c.s.)

The linear correlation of centre shift with sum of electronegativities (Fig. 5.3) shows that the centre shift values could be expressed by the algebraic sum of an empirical parameter for each ligand which we call a partial centre shift (p.c.s.):

$$\text{C.S.} = \sum_{i=1}^{6} (\text{p.c.s.})_i$$

Using the above equation for the $SnX_4Y_2^=$ series, we derive $(\text{p.c.s.})_{Cl^-} = 0.09$, $(\text{p.c.s.})_{Br^-} = 0.14$, $(\text{p.c.s.})_{I^-} = 0.21$ and $(\text{p.c.s.})_{F^-} = -0.04$ (all in mm s^{-1}) from the centre shifts (Table 5.6) of the hexahalides and $SnCl_4F_2^=$. These values may then be used to predict the centre shift values for other compounds in Table 5.6. For $SnCl_4Br_2^=$ and $SnI_4Br_2^=$, the predicted centre shift values using the above partial centre shift values and above equation are 0.64 mm s^{-1} and 1.12 mm s^{-1} respectively, compared to the observed values of 0.65 mm s^{-1} and 1.11 mm s^{-1}. Thus, for this series of compounds, the partial centre shift values would be useful for predicting centre shift values for new compounds such as the mixed fluorides, and perhaps for distinguishing between *cis* and *trans* isomers: i.e. *cis* isomers might be expected to have slightly different centre shift values than those predicted from partial centre shift values derived from *trans* structures. In larger series of compounds such as the Fe[II] low spin series discussed later, partial centre shift values are very valuable for unravelling trends in centre shift data, and for systematizing and correlating centre shift values.

The concept of partial centre shifts was first introduced (ref. 5.3) for a large series of organometallic compounds. The additivity theory was tested for low oxidation state Fe in five- and six-coordination, and a good degree of internal consistency was obtained. However, it has become apparent that like electronegativity, one partial centre shift value for a ligand would not be expected to hold for one Mössbauer element in all its coordination numbers and oxidation states. Separate values have to be derived for each Mössbauer ion in a given fixed coordination number: for example, Fe^{II} low spin in six-coordination or Sn^{IV} in four-coordination. Even then, partial centre shift values cannot always be derived, and several other assumptions inherent in the initial postulate should be emphasized. A constant partial centre shift value implies that the centre shift is insensitive to any small changes in bond angles and metal–ligand bond distances which occur from compound to compound, and also implies that the centre shift depends only on the ligands directly bonded to the Mössbauer atom. Also the partial centre shift value for one ligand must be relatively insensitive to the bonding properties of the other ligands in the compounds. Thus, ideally the variation in partial centre shift value of one ligand should be small in comparison with the differences between partial centre shift values from one ligand to another. The partial centre shift value of a strong π acceptor ligand such as CO or NO^+ would be expected to change markedly from compound to compound, just as the infrared stretching frequency for these ligands changes markedly from compound to compound.

Besides the $SnX_4Y_2^=$ series mentioned above, partial centre shift values can be derived for other series of compounds. For Fe^{2+} high spin compounds, a plot of isomer shift (centre shift – second order Döppler shift) versus n for $[FeCl_{6-n}\cdot nH_2O]^{n-4}$ species produces a straight line (Fig. 5.6) implying that partial centre shift values could be derived for H_2O and Cl. Some fortuitous cancellation of effects is probably operating here because the Fe–Cl and Fe–O bond lengths vary markedly from compound to compound.

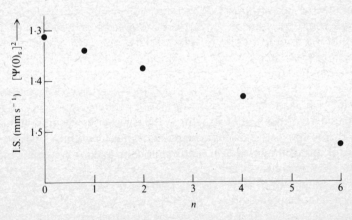

Fig. 5.6 Isomer shift (with respect to natural iron at room temperature) versus hydration number n for the series $[FeCl_{6-n}\cdot nH_2O]^{n-4}$. (Y. Hazony, R. C. Axtmann, and J. W. Hurley, *Chem. Phys. Lett.*, **2**, 440 (1968).)

Mössbauer spectroscopy

Table 5.10   Centre shifts (relative to AuPt source) for gold compounds (mm s^{-1} at 4·2 K)

Compound	C.S.	Compound	C.S.
KAuI$_4$	+0·43	As(C$_6$H$_5$)$_4$Au(N$_3$)$_4$	+1·66
KAuBr$_4$	+0·60	Na$_3$AuO$_3$	+2·45
KAuCl$_4$	+0·81	KAu(CN)$_2$Br$_2$	+2·65
KAu(SCN)$_4$	+1·63	KAu(CN)$_4$	+4·03

H. D. Bartunik, W. Potzel, R. L. Mössbauer and G. Kaindl, *Z. Physik*, **240**, 1 (1970).

Similarly, results for Au^{III} compounds also indicate that partial centre shift values could be derived (Table 5.10). Thus [Au(CN)$_2$Br$_2$]$^-$ has a centre shift value (2·65 mm s^{-1}) close to the arithmetic mean of the centre shift values for AuBr$_4^-$ and Au(CN)$_4^-$ (2·32 mm s^{-1}) as expected if the centre shift is an algebraic sum of partial centre shift values.

Only for Fe^{II} and Sn^{IV} compounds has the partial centre shift concept been widely tested. For Fe^{II} low spin compounds, self-consistent partial centre shift values have been derived for a large number of ligands. Table 5.11 gives a small number of Fe^{II} low spin centre shift results, from which the partial centre shift values in Table 5.12 were derived using the equation:

$$\text{C.S.} = 0.16 + \sum_{i=1}^{6} (\text{p.c.s.})_i \qquad (5.6)$$

and arbitrarily taking the partial centre shift value for RNC (R = Me, Et, p-MeO–C$_6$H$_4$·NC) to be zero. From compounds 3 and 4 (Table 5.11), the values of Cl$^-$ and SnCl$_3^-$ are derived taking (p.c.s.)$_{ArNC}$ = 0. Similarly, the partial centre shift value of depe is derived from compound 5 taking the (p.c.s.)$_{Cl^-}$ = 0·10 from compound 3. In this way, the partial centre shift values in Table 5.12 were derived using *trans* isomers wherever possible. The predicted and observed values for a number of cross-check compounds are given in the second part of Table 5.11 (compounds 11–19). Except for compound 11, the agreement between predicted and observed values is good, and the initial assumptions are strongly supported for the majority of these ligands. In particular, the predicted and observed value for *trans*-FeHCl(depe)$_2$ (compound 14) is in good agreement with the observed value. This indicates that the bonding properties of strong σ donors such as H$^-$ do not change markedly from *trans*-FeH$_2$(depb)$_2$ to *trans*-FeHCl(depe)$_2$. However, the observed centre shift value for *cis*-FeCl$_2$(ArNC)$_4$ (compound 11) is appreciably different than that for *trans*-FeCl$_2$(ArNC)$_4$ (compound 3). This indicates that in the *cis* compound, the π acceptor capability of ArNC increases when *trans* to a weak π acceptor such as Cl$^-$, thus increasing [Ψ(0)$_s$]^2 and decreasing

Table 5.11 Centre shift values (relative to nitroprusside at 295 K) for Fe^{II} low spin compounds (mm s^{-1})

Compound	C.S.	p.c.s. *obtained*	Compound	C.S. Found	C.S. Predicted
1. Fe(MeNC)$_6$(HSO$_4$)$_2$	+0.14	MeNC = 0.00	11. *cis*-FeCl$_2$(ArNC)$_4$	+0.28	+0.36
2. Fe(EtNC)$_6$(ClO$_4$)$_2$	+0.16	EtNC = 0.00	12. *cis*-Fe(SnCl$_3$)$_2$(ArNC)$_4$	+0.27	+0.24
3. *trans*-FeCl$_2$(ArNC)$_4$	+0.36	Cl$^-$ = +0.10	13. [FeCl(ArNC)$_5$]ClO$_4$	+0.23	+0.26
4. *trans*-Fe(SnCl$_3$)$_2$(ArNC)$_4$	+0.24	SnCl$_3^-$ = +0.04	14. *trans*-FeHCl(depe)$_2$	+0.39	+0.42
5. *trans*-FeCl$_2$(depe)$_2$	+0.59	depe/2 = +0.06	15. *trans*-FeCl(SnCl$_3$)(depe)$_2$	+0.55	+0.54
6. *trans*-FeCl$_2$(depb)$_2$	+0.59	depb/2 = +0.06	16. *trans*-Fe(CN)$_2$(MeNC)$_4$	+0.16	+0.18
7. *trans*-FeH$_2$(depb)$_2$	+0.23	H$^-$ = −0.08	17. *cis*-Fe(CN)$_2$(MeNC)$_4$	+0.16	+0.18
8. K$_4$Fe(CN)$_6$	+0.21	CN$^-$ = +0.01	18. *trans*-Fe(CN)$_2$(EtNC)$_4$	+0.21	+0.18
9. K$_2$[Fe(niox)$_2$(CN)$_2$]	+0.34	niox/2 = +0.04	19. Na$_3$[Fe(CN)$_5$NH$_3$]·H$_2$O	+0.26	+0.28
10. Fe(niox)$_2$(NH$_3$)$_2$	+0.46	NH$_3$ = +0.07			

ArNC = p-methoxyphenylisocyanide; depe = 1,2 bis(diethylphosphine)-ethane; depb = o-phenylenebisdiethylphosphine; niox = 1,2-cyclohexanedione dioxime.

G. M. Bancroft, M. J. Mays and B. E. Prater, *J. Chem. Soc. (A)*, 956 (1970).

the centre shift from that of the *trans* compound. Stronger π acceptors such as CO give more marked deviations, and indicate that a partial centre shift value cannot be realistically assigned to such a ligand. For example, the predicted and observed centre shift value for *trans*-$[FeH(CO)(depe)_2]^+BPh_4^-$ (Table 5.3) are $+0.32$ mm s^{-1} and $+0.12$ mm s^{-1} respectively, and the very low CO stretching frequency (1915 cm^{-1}) indicates that CO is acting as an even stronger π acceptor than usual, decreasing the centre shift from the expected value. Similarly, for other compounds such as $Fe(CO)_2X_2P_2$ and $Fe(CO)_3X_2P$ (X = Cl$^-$, Br$^-$, I$^-$; P = PPh$_2$Me, PPh$_3$), predicted and observed values are not in agreement. Despite these discrepancies, it appears that except for strong π acceptors such as CO or NO$^+$, partial centre shift values for ligands can be derived, remain reasonably constant from compound to compound, and provide a reasonably accurate guide to Fe^{II} low spin centre shift values.

The partial centre shift values, like centre shift values for Fe^{II} compounds, decrease with increasing σ bonding and π backbonding:

$$\text{p.c.s.} = -k(\sigma + \pi) \qquad (5.7)$$

Along with the partial quadrupole splittings derived in the next chapter, σ and π can at least be separated qualitatively. Consistent with the above equation, the best π acceptor (NO$^+$) and the best σ donor (H$^-$) give the most negative partial centre shift values (Table 5.12), while ligands giving

Table 5.12 Partial centre shifts (mm s^{-1})†

	p.c.s.		p.c.s.		p.c.s.
NO$^+$	−0.20*	SnCl$_3^-$	0.04	phen/2	0.07
		qp/4	0.05		
H$^-$	−0.08	niox/2	0.04	py	0.07
SiH$_3^-$	−0.05	SbPh$_3$	0.05*	NH$_3$	0.07
PhCH$_2$NC	−0.01	NO$_2^-$	0.05*	but	0.07
ArNC	0.00	pc/4	0.05	AsPh$_3$	0.08*
MeNC	0.00	dmpe/2	0.05	pip	0.08
EtNC	0.00	NCS$^-$	0.05	pyim	0.08
CO*	0.00	NCO$^-$	0.06	tripyam/3	0.08
CN$^-$	0.01	depe/2	0.06	Im	0.08
SO$_3^{2-}$	0.01*	depb/2	0.06	N$_3^-$	0.08
P(C$_6$H$_5$)$_3$	0.02*	bipy/2	0.06	H$_2$O	0.10*
				Cl$^-$	0.10
				Br$^-$	0.13
				I$^-$	0.13

* This p.c.s. is probably significantly lower than this, and quite variable.
† Estimated error = ± 0.01 mm s^{-1} except for those marked with an asterisk (*) for which the estimated error is $\geq \pm 0.02$ mm s^{-1}.
(G. M. Bancroft, M. J. Mays, and B. E. Prater, *J. Chem. Soc.* (A), 956, (1970)).

the most positive partial centre shift values (I^-, Br^-, Cl^-) give the most ionic bonds with comparatively little covalent character.

For a limited number of Sn^{IV} compounds such as the $SnX_4Y_2^=$ series discussed earlier, self-consistent partial centre shift values can be derived. However, for many compounds such as the tin organometallics mentioned earlier (Table 5.9), partial centre shift values cannot be derived. For example, in the R_nSnX_{4-n} series (ref. 5.1), the centre shift would be expected to increase regularly with increasing n. Instead, the centre shift reaches a maximum at $n = 2$ before levelling off or dropping.

As more centre shift data become available for such isotopes as ^{99}Ru, ^{197}Au, and ^{193}Ir, it will be interesting to see whether the partial centre shift concept becomes more widely useful. Because Ru and Ir compounds often contain strong π acceptor ligands (Table 5.4), such as CO and NO^+, partial centre shift values for these species may be of limited value.

5.3 Correlation of centre shift values with the spectrochemical and nephelauxetic series

For Fe^{II} low spin, Ru^{II}, Ir^{III} and Au compounds, correlations between the centre shift and the spectrochemical ranking of ligands have been observed; while for Fe^{II} high spin compounds, the centre shift has been correlated with the nephelauxetic series of ligands in which the ligands are arranged according to decreasing Racah parameters (decreasing interelectronic repulsion in the complex). The former correlation with the spectrochemical series is expected for more covalent compounds, while the correlation with the nephelauxetic series is expected for less covalent compounds in which covalency is determined in the Mulliken approximation by the differences in orbital electronegativities between metal and ligand (ref. 5.4).

The correlation of partial centre shift values for the ligands in Table 5.12 with the partial ligand field strength is shown in Fig. 5.7. This, and the following correlations are perhaps not too surprising because just as $[\Psi(0)_s]^2$ should increase (centre shift decrease for ^{57}Fe) with an increase in $(\sigma + \pi)$ so the ligand field strength Δ values should increase i.e.

$$\Delta = \Delta \langle V_{oct} \rangle + \sigma' + \pi' \qquad (5.8)$$

$\Delta \langle V_{oct} \rangle$, the crystal field term is usually considered to be small, and hence Δ should increase with increasing $\sigma + \pi$. Thus NO^+, the strongest field ligand gives rise to the lowest (most negative) partial centre shift, while Br^- and I^-, the weakest field ligands give the most positive partial centre shift values. This correlation indicates that H^- occupies a position in the spectrochemical series near or above CN^-, and not between H_2O and NH_3 as had been previously suggested. Similarly, for the $[Ru(X)_5NO]^=$ compounds, (Table 5.4) the centre shift values increase in the order $Br^- < Cl^- < NCS^- < NH_3 < CN^-$; and for the Au^{III} compounds (Table 5.10) the centre shift

Mössbauer spectroscopy

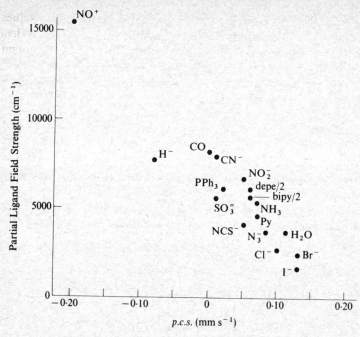

Fig. 5.7 Partial centre shifts for Fe^{II} low spin ligands versus partial ligand field strength (δ). (G. M. Bancroft, M. J. Mays, and B. E. Prater, *J. Chem. Soc. (A)*, 956 (1970).)

increases in the order $I^- < Br^- < Cl^- < NCS^- \sim N_3^- < O^= \ll CN^-$. Both of these series correlate well with the spectrochemical ranking of ligands.

By contrast, for less covalent compounds such as the high spin ferrous halides (Fig. 5.5), the centre shift decreases ($[\Psi(0)_s]^2$ increases) in the order $F^- > Cl^- > Br^- > I^-$ i.e. in the order of electronegativities or the nephelauxetic series.

Problems

1. How would you expect the centre shift to differ in the following pairs of compounds? Give your reasoning.
   (a) $FeBr_3$ and $FeBr_2$
   (b) $SnBr_4$ and $SnBr_2$
   (c) $AuBr_3$ and $AuBr$
   (d) $SbBr_5$ and $SbBr_3$
   (e) $Fe(CO)_5$ and $Fe(CO)_4PPh_3$

(f) *trans*-[FeH(N$_2$)(dppe)$_2$]$^+$ and *trans*-[FeH((CH$_3$)$_2$CO)(dppe)$_2$]$^+$
(dppe = Ph$_2$PCH$_2$CH$_2$PPh$_2$)
(g) Ph$_3$SnMn(CO)$_5$ and Cl$_3$SnMn(CO)$_5$
2. Using the partial centre shifts in Table 5.12, calculate the expected centre shifts for the following Fe^{II} low spin compounds at room temperature.
   (a) *trans*-Fe(NCS)$_2$(depe)$_2$
   (b) *trans*-FeClSnCl$_3$(ArNC)$_4$
   (c) *cis*-Fe(NCS)$_2$(depe)$_2$
   (d) *trans*-[FeH(NH$_3$)(depe)$_2$]$^+$
   (e) *trans*-FeH$_2$(depb)$_2$
3. *Cis*-octahedral Sn^{IV} compounds invariably give substantially lower centre shift values than corresponding *trans*-octahedral isomers. For example, the centre shifts of *cis*-Ph$_2$Sn(acac)$_2$ and *trans*-Me$_2$Sn(acac)$_2$ are 0·74 mm s^{-1} and 1·18 mm s^{-1} respectively. Suggest a possible explanation.

References

5.1 R. V. Parish, in *Prog. Inorganic Chemistry*, **15**, 101 (1972).
5.2 G. K. Shenoy and S. L. Ruby, *Mössbauer Effect Methodology*, vol. 5 ed. I. J. Gruverman, Plenum Press (1970), p. 77.
5.3 R. H. Herber, R. B. King and G. K. Wertheim, *Inorg. Chem.*, **3**, 101 (1964).
5.4 H. D. Bartunik, W. Potzel, R. L. Mössbauer, and G. Kaindl, *Z. Physik*, **240**, 1 (1970).

6. Bonding and structure from quadrupole splittings

As discussed in chapter 2, the degeneracy of nuclear energy levels is removed by the interaction of an electric field gradient with a nuclear quadrupole moment. For ^{57}Fe this interaction results in a typical two-line Mössbauer spectrum (Fig. 2.3). For convenience, the field gradient q was divided into a $q_{valence}$ term and a $q_{lattice}$ term (eq. 2.12), and $q_{valence}$ was further subdivided into $q_{C.F.}$ which arises from a d-orbital imbalance due to crystal field splitting, and $q_{M.O.}$, which arises from d- or p-orbital imbalance due to covalent bonding (eq. 2.16). For example, considering just p-orbital populations, a $q_{M.O.}$ term arises if $N_{p_z} \neq \frac{1}{2}(N_{p_x} + N_{p_y})$.

We begin this chapter by correlating quadrupole-splitting trends due to variations in $q_{M.O.}$ with changes in bonding properties of the ligands. For each series of compounds, the Z EFG axis is defined (usually the molecular axis of highest symmetry). We then use the ideas outlined in chapter 2.2 to discuss changes in quadrupole splittings. For positive Q, replacement of a poor σ donor (an electronegative ligand such as Cl) along the Z EFG axis by a good σ donor (electropositive ligand such as H or CH$_3$) will lead to a concentration of negative charge along the Z EFG axis, and give a more negative q and quadrupole splitting. For example, considering I$_2$ and ICl, we choose the Z EFG axis through the I–L bond axis. The more electronegative Cl will decrease N_{p_z} on the iodine atom in ICl relative to N_{p_z} in I$_2$. Thus, q will be more *positive* in ICl than I$_2$. Just as $\delta R/R$ can be positive or negative, Q can be positive or negative. Since Q is negative for ^{129}I, ICl has a more negative quadrupole splitting than I$_2$.

If ligand substitution along the X or Y EFG axes leads to an electron concentration in p_x and p_y, then q and the quadrupole splitting (for positive Q) becomes more positive. For example, for cis-MA_2B_4 compounds (Fig. 2.5), if the A ligands along the X and Y axes were changed from the comparatively weak σ donor Cl$^-$ to the strong σ donor CN$^-$, then the quadrupole splitting will become more *positive* (for +ve Q). If Q were negative, then substitution of CN$^-$ for Cl$^-$ would give a more *negative* quadrupole splitting.

Bonding and structure from quadrupole splittings

The above ideas are used to rationalize quadrupole splitting trends in I, Xe, Fe, and Ru compounds in which just one or two ligands are varied. Correlations of centre shifts with quadrupole splittings are widely used in the first section of this chapter to examine the relative sensitivity of the quadrupole splittings to σ donor and π acceptor ligands. Such correlations are very useful in qualitatively separating σ donor and π acceptor properties of ligands such as dinitrogen. This approach is useful for a large number of other Mössbauer ions.

In addition to bonding information, quadrupole splittings have recently been widely and usefully used for structural predictions. These predictions are especially important for Sn^{IV} compounds, because many compounds with four-coordinate stoichiometry in fact possess associated structures in the solid state involving five- and six-coordinate tin atoms. For example, Me_3SnCl apparently involves a five-coordinate Sn atom, while Ph_3SnCl is four-coordinate. The structural predictions from Mössbauer quadrupole splittings are usually based on derived partial quadrupole splittings using the point charge expressions for the EFG components (Table 2.2). The assumptions involved in using the partial quadrupole treatment will be emphasized in the second section of this chapter. Several examples of its use in structural determinations for compounds of Fe^{II}, Sn^{IV}, Ru^{II}, Ir^{III}, and Co^{III} will be discussed. The partial quadrupole splitting values are also extremely useful in systematizing quadrupole splitting data and in obtaining semi-quantitative bonding information for a host of ligands.

Finally, in the last section of this chapter, we will discuss the effect of $q_{C.F.}$ on quadrupole splittings. For Fe^{2+} high spin, the $q_{C.F.}$ term dominates the quadrupole splittings. A measurement of the quadrupole splitting as a function of temperature enables any distortion from octahedral or tetrahedral symmetry to be detected, and an estimation of the t_{2g} crystal field splittings. In addition, the quadrupole splittings can be correlated with estimates of $3d$-electron delocalization. For other Mössbauer ions such as Fe^{0} and Au^{III} (both having d^8 electronic configurations), both $q_{M.O.}$ and $q_{C.F.}$ are important in determining the quadrupole splittings. The separation of $q_{valence}$ into $q_{C.F.}$ and $q_{M.O.}$ is useful in discussing Au^{III} quadrupole splitting trends, but provides little, if any, illumination on the trends in Fe^{0} quadrupole splittings.

6.1 Correlation of quadrupole splittings with bonding properties of ligands and centre shifts

Iodine compounds. Iodine quadrupole splittings provide the simplest examples of the Townes–Dailey approach described in chapter 2.2.† The

† For a more detailed approach to the iodine literature, see reference 6.1.

Mössbauer spectroscopy

molecular quadrupole splitting can be written using the $q_{valence}$ expression from eq. 2.14:

$$eQV_{ZZ} = e^2qQ = (1 - R)e^2QK_pU_p \quad (6.1)$$

where:

$$U_p = [-N_{p_z} + \tfrac{1}{2}(N_{p_x} + N_{p_y})] \quad (6.1a)$$

For atomic iodine ($5p^5$), $U_p = 1$ and $eQV_{ZZ} = -2293$ Mc s^{-1} (see Appendix 3). Thus $(1 - R)e^2QK_p$ is -2293 Mc s^{-1} and eq. (6.1) can now be written:

$$eQV_{ZZ} = -2293\, U_p \quad (6.2)$$

Using eq. (6.2), U_p can then be calculated for any iodine compound if eQV_{ZZ} can be measured. The other components of the EFG can be written by analogous equations:

$$eQV_{XX} = -2293[-N_{p_x} + \tfrac{1}{2}(N_{p_y} + N_{p_z})] \quad (6.3)$$
$$eQV_{YY} = -2293[-N_{p_y} + \tfrac{1}{2}(N_{p_x} + N_{p_z})] \quad (6.4)$$

Since

$$\eta = \frac{V_{XX} - V_{YY}}{V_{ZZ}}$$

substitution of eqs. (6.2), (6.3) and (6.4) gives

$$\eta = \frac{\tfrac{3}{2}(N_{p_y} - N_{p_x})}{U_p} \quad (6.5)$$

Using eqs. (6.2), (6.5) and the expression derived previously for the centre shift (eq. 5.1), it is possible to determine N_{p_x}, N_{p_y}, and N_{p_z} from a Mössbauer spectrum assuming that $q_{lattice}$ is negligible, and that the intermolecular influence on the charge density at the iodine nucleus can be neglected. These two assumptions may not always be valid, but it is apparent that the centre shift and quadrupole splitting for iodine compounds provide a powerful means for determining the covalent character of iodine bonds.

A few examples illustrate the above discussion. Consider IBr, ICl, and SnI$_4$(Table 6.1). The Z EFG axis is chosen along the bond axis. N_{p_z} will then increase or decrease from the atomic iodine value of one via σ bonding. For IBr or ICl, we would expect that N_{p_z} would be substantially less than one, since chlorine and bromine are substantially more electronegative than iodine. N_{p_y} and N_{p_x} will decrease from the atomic I value of two by π bonding. Summarizing:

$$N_{p_z} = 1 \pm \sigma; \quad N_{p_y} = 2 - \pi_y; \quad N_{p_x} = 2 - \pi_x \quad (6.6)$$

Consider ICl. The quadrupole splitting, η and h_p values are given in Table 6.1.

Table 6.1 5p orbital populations in iodine compounds from C.S. and Q.S. at ~80 K

Compound	$e^2qQ(^{127}\text{I})$† Mc s^{-1}	η	h_p^*	U_p‡	N_{p_x}	N_{p_y}	N_{p_z}
IBr	−2892	0.06	1.30	1.26	1.98	2.00	0.72
ICl	−3131	0.06	1.67	1.37	1.87	1.92	0.53
SnI$_4$	−1364	0	0.71	0.60	1.96	1.96	1.37

† For ^{127}I, Q is negative. $e^2qQ(^{127}\text{I}) = (Q_{127}/Q_{129})\,e^2qQ(^{129}\text{I})$. For the conversion of velocities to Mc s^{-1}, see Appendix 3.
* From centre shift data and eq. 5.1 (Table 5.1).
‡ $U_p = e^2qQ/-2293$ (eq. 6.2).
M. Pasternak and T. Sonnino, *J. Chem. Phys.*, **48**, 1997 (1968); S. Bukshpan and R. H. Herber, *J. Chem. Phys.*, **46**, 3375 (1967).

Using the eqs. (5.1), (6.5) and (6.2), we can then write:

$$1.67 = 6 - (N_{p_z} + N_{p_x} + N_{p_y}) \quad \text{(from 5.1)}, \qquad (6.7\text{a})$$

$$0.05 = N_{p_y} - N_{p_x} \quad \text{(from 6.5)}, \qquad (6.7\text{b})$$

$$1.37 = -N_{p_z} + \tfrac{1}{2}(N_{p_x} + N_{p_y}) \quad \text{(from 6.2)}. \qquad (6.7\text{c})$$

Solving these three equations, the values for $N_{p_x}, N_{p_y},$ and N_{p_z} are readily calculated (Table 6.1). Similarly, from the quadrupole splittings and centre shifts for IBr and SnI$_4$, the orbital populations shown in Table 6.1 can be calculated. Since the iodine in the alkali halides approaches the electronic configuration $5p^6$, $N_{p_z} \approx \tfrac{1}{2}(N_{p_x} + N_{p_y})$ ($U_p \rightarrow 0$), and no quadrupole splitting is resolved. One slightly broadened line is observed.

Because the electronegativities of the ligands vary in the order Cl > Br > Sn, we would expect that N_{p_z} would decrease in the order SnI$_4$ > IBr > ICl with $\tfrac{1}{2}(N_{p_x} + N_{p_y})$ remaining fairly constant near 2 because of little or no π bonding (Table 6.1). The slightly smaller values for ICl ($N_{p_x} = 1.92$, $N_{p_y} = 1.87$) may be due to a breakdown in the assumptions, such as the assumed linear relationship of the centre shift with h_p (Fig. 5.1), or a significant intermolecular association. U_p becomes more positive in the order SnI$_4$ < IBr < ICl; since Q is negative for ^{129}I, the quadrupole splittings become more negative in the above order.

A plot of centre shift versus quadrupole splitting is shown for the above mentioned compounds in Fig. 6.1. The s-electron density at the nucleus decreases (as given by a decrease in centre shift) as U_p decreases (as given by an increasing positive quadrupole splitting). Remember that $\delta R/R$ and Q are positive and negative respectively for ^{129}I. This correlation is probably expected since our previous discussion indicated that both the centre shift and quadrupole splitting were sensitive to changes in 5p-electron density—mostly to changes in N_{p_z}. Thus as the electronegativity of the ligand decreases,

Mössbauer spectroscopy

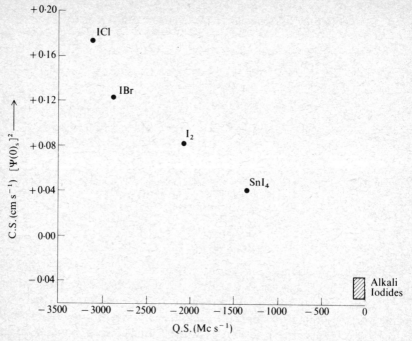

Fig. 6.1 Centre shift versus quadrupole splitting for iodine compounds (references below Tables 5.1 and 6.1).

the $5p$ electron population on iodine increases, giving a decreasing $[\Psi(0)_s]^2$ and a decreasing U_p.

Several other plots of centre shift versus quadrupole splitting will be shown in the next few sections, and generally a good correlation is obtained. Since the signs of $\delta R/R$ and Q can lead to confusion in interpreting changes in electron populations due to changes in bonding properties of ligands, we always plot the centre shift such that $[\Psi(0)_s]^2$ increases along the Y axis, and we plot the quadrupole splitting such that the σ donor capacity of the varied ligands increases (or generally the electronegativity decreases) along the X axis.

The slope of the line in Fig. 6.1 is negative. Similarly for xenon, the centre shift versus quadrupole splitting plot (Fig. 6.2) gives a negative slope whereas for other ions, the centre shift versus quadrupole splitting plot gives positive slopes (Figs. 6.3, 6.4, 6.5). The difference in slopes is associated with the change of the centre shift due to p orbital occupancy changes in I and Xe, but s orbital occupancy changes in Fe, Ru and Sn (chapter 5).

Xenon compounds. The quadrupole splittings in Table 6.2 for xenon tetrahalides and dihalides again can be rationalized by considering the $5p$ orbital imbalance on xenon created by ligands of varying electronegativity.

Bonding and structure from quadrupole splittings

Table 6.2 Q.S. values for xenon compounds at 4·2 K

Substance*	$\tfrac{1}{2}e^2qQ$ (mm s^{-1})	e^2qQ† (Mc s^{-1})	$\lvert U_p \rvert$
XeF$_4$	41·04	(+) 2620	1·50
[XeCl$_4$]	25·62	(+) 1640	0·94
XeF$_2$	39·00	(−) 2490	1·43
[XeCl$_2$]	28·20	(−) 1800	1·03
[XeBr$_2$]	22·2	(−) 1415	0·81

* The compounds in brackets were 'prepared' from the corresponding iodine compound and used as a source. For example, XeCl$_4$ was 'prepared' from KICl$_4$.

† e^2qQ for one p_z electron in Xe is 1740 Mc s^{-1} or 54·6 mm s^{-1}, taking E_γ = 39·58 keV (Appendix 3).

G. J. Perlow and H. Yoshida, *J. Chem. Phys.*, **49**, 1474 (1968).

The dihalides have a linear structure, and we take the Z EFG axis through the molecular bond axis; the tetrahalides are square planar and we take the Z EFG axis through the four fold molecular axis—again the highest symmetry axis. Assuming that only Xe p electrons partake in the bonding (as in section 5.1), just the p_z electrons will be utilized in σ bonding in the dihalides, while $5p_x$ and $5p_y$ electrons partake in σ bonding in the tetrahalides.

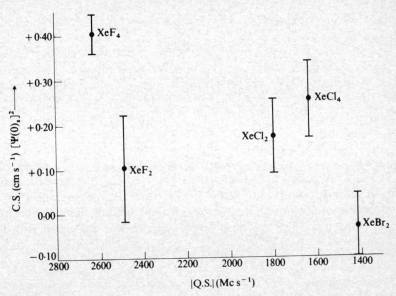

Fig. 6.2 Centre shift versus quadrupole splitting for xenon compounds (Tables 5.2 and 6.2).

Mössbauer spectroscopy

Neglecting π bonding N_{p_x} and N_{p_y} should be very close to two in the dihalides, while N_{p_z} should be close to two in the tetrahalides. Thus in the dihalides, $N_{p_z} < \frac{1}{2}(N_{p_x} + N_{p_y})$ and U_p is positive; while in the tetrahalides, $N_{p_z} > \frac{1}{2}(N_{p_x} + N_{p_y})$ and U_p is negative. Since Q is negative for ^{129}Xe, the quadrupole splitting will be negative for the dihalides and positive for the tetrahalides. For both series of compounds $|U_p|$ increases as the electronegativity of the ligands increases. Thus, for the dihalides, the quadrupole splitting becomes more negative in the order $XeBr_2 < XeCl_2 < XeF_2$; while for the tetrahalides, the quadrupole splitting becomes more positive in the same order. Since it is known from spectroscopic and atomic beam measurements that e^2qQ for one p_z electron in Xe is 1742 Mc s^{-1} ($= 54.6$ mm s^{-1}), then the values of $|U_p|$ in Table 6.2 can be calculated using eq. (6.1).

A plot of centre shift versus quadrupole splitting (Fig. 6.2) shows that the centre shift and $[\Psi(0)_s]^2$ increase generally (considering the dihalides and tetrahalides separately) as the magnitude of the quadrupole splitting increases: i.e., $[\Psi(0)_s]^2$ increases as the 5p orbital occupancy decreases and $|U_p|$ increases.

Fe^{II} and Ru^{II} compounds. As discussed in section 5.1, bonding in Fe^{II} and Ru^{II} compounds consists of σ donation from the ligands to the metal $d^2 sp^3$ hybrid, and π back donation from the metal t_{2g} orbitals to the empty π orbitals. We showed in chapter 5 that both σ donation and π acceptance increased $[\Psi(0)_s]^2$ in Fe^{II} and Ru^{II} compounds, decreasing the centre shift for Fe^{II} ($\delta R/R = -$ve), but increasing the centre shift for Ru^{II} ($\delta R/R = +$ve). Thus in the trans-[Fe^{II}HL(depe)$_2$]$^+$BPh$_4^-$ series, ($\sigma + \pi$) increased in the order RCN < N$_2$ < P(OPh)$_3$ < P(OMe)$_3$ < Me$_3$CNC < p−MeO.C$_6$H$_4$.NC < CO.

The above qualitative molecular orbital treatment for quadrupole splittings leads to a qualitative separation of σ and π for the above ligands, and a useful explanation of quadrupole splitting trends in a number of series of compounds. Before discussing the quadrupole splittings for Fe^{II} compounds (Table 6.3), the signs of at least some of these quadrupole splittings must be known. In the [Fe(CN)$_5$X$^{n-}]^{(3+n)-}$ series, the quadrupole splitting (and q) has been found to be positive for Na$_2$Fe(CN)$_5$NO.2H$_2$O, Na$_3$Fe(CN)$_5$NH$_3$.H$_2$O and K$_3$Fe(CN)$_5$H$_2$O using the magnetic field technique (section 2.3). Based on evidence presented later in this chapter, it seems reasonable that the other compounds of the [Fe(CN)$_5$X$^{n-}]^{(3+n)-}$ series also have positive values. By contrast, trans-[FeH(pMeO.C$_6$H$_4$.NC)(depe)$_2$]$^+$BPh$_4^-$ has a negative quadrupole splitting (and q), and later evidence will strongly suggest that the other compounds in this series have negative quadrupole splittings.

For both series of compounds, the Z EFG axis is taken through the highest symmetry axes in the ion—in both cases a four-fold axis. This choice

Bonding and structure from quadrupole splittings

Table 6.3 Quadrupole splittings for $(Fe(CN)_5X^{n-})^{(3+n)-}$ and trans-$[FeHL(depe)_2]^+BPh_4^-$ compounds (mm s^{-1} at room temperature)

X	Q.S.*	L	Q.S.
NO$^+$	+1·73	CO	(−)1·00
CO	(+)0·43	pMeO.C$_6$H$_4$.NC	−1·14
SO$_3^=$	(+)0·80	Me$_3$CNC	(−)1·13
P(C$_6$H$_5$)$_3$	(+)0·62	P(OMe)$_3$	(−)0·92
NO$_2^-$	(+)0·89	P(OPh)$_3$	(−)0·72
Sb(C$_6$H$_5$)$_3$	(+)0·94	N$_2$	(−)0·33
NH$_3$	+0·67	PhCN	(−)0·58
As(C$_6$H$_5$)$_3$	(+)0·92	MeCN	(−)0·46
H$_2$O	+0·80	Cl	≤0·12
		I	≤0·19

* The signs in brackets are predicted signs (see text).
E. Fluck, W. Kerler, and W. Neuwirth, *Angew. Chem. Internat. Edn.*, **2**, 277 (1963); N. L. Costa, J. Danon and R. M. Xavier, *J. Phys. Chem. Solids*, **23**, 1783 (1962); E. Fluck and P. Kuhn, *Z. anorg. Chem.* **350**, 263 (1967); N. E. Erickson, Ph.D. Thesis, University of Seattle, Washington, U.S.A. (1964); G. M. Bancroft, M. J. Mays, B. E. Prater, and F. P. Stefanini, *J. Chem. Soc.* (A), 2146 (1970).

makes the X and Y axes equivalent (for ideal geometry), and η should be zero. The ligands X (in the Fe(CN)$_5$X series) and L (in the trans-$[FeHL(depe)_2]^+$ series) both lie along the Z EFG axis.

We consider that any quadrupole splitting is due mainly to a 3d orbital imbalance (U_d); and as in the centre shift treatment, that any change in U_d is mainly due to the varied ligand and not due to a change in bonding properties of the fixed ligands. We also neglect intermolecular contributions, or any $q_{lattice}$ effects from cations or anions. For the Fe^{II} cyanide series, if X = CN$^-$, then the electronic and ligand symmetry about the iron atom is cubic and no quadrupole splitting is observed. If X is a better σ donor than CN$^-$, then a negative contribution to $q_{M.O.}$ is expected ($N_{d_{z^2}} > N_{d_{x^2-y^2}}$) (eq. 2.15); while if X is a better π acceptor than CN$^-$, then charge is delocalized from the Z EFG axis via d_{xz} and d_{yz}, and a positive contribution to $q_{M.O.}$ is expected $[N_{d_{xy}} > \frac{1}{2}(N_{d_{xz}} + N_{d_{yz}})]$. The sign of q (and the quadrupole splitting) will depend on *both* the σ donor and π acceptor strengths *relative* to CN$^-$, and will become more *negative* with increasing ($\sigma - \pi$). The positive quadrupole splitting signs for these compounds indicate that all ligands are poorer σ donors and/or better π acceptors than CN$^-$, i.e., poorer ($\sigma - \pi$) ligands. The magnitude of the quadrupole splitting values (Table 6.3) indicates that σ *increases* and/or π *decreases* in the order NO$^+$ < Sb(C$_6$H$_5$)$_3$ < As(C$_6$H$_5$)$_3$ < NO$_2^-$ < SO$_3^=$ ~ H$_2$O < P(C$_6$H$_5$)$_3$ < CO. Thus NO$^+$, the strongest π acceptor gives the most positive quadrupole splitting.

Mössbauer spectroscopy

The trans-[FeHL(depe)$_2$]$^+$BPh$_4^-$ quadrupole splittings (Table 6.3) lead to a qualitative separation of σ and π effects for molecular N$_2$ in conjunction with centre shift data for these compounds (Table 5.3). As previously discussed for the Fe^{II} cyanides, we would expect the quadrupole splitting to become more negative as ($\sigma - \pi$) for L increases. The quadrupole splittings show that $\sigma - \pi$ increases in the order N$_2$ < MeCN < PhCN < P(OPh)$_3$ < P(OMe)$_3$ < CO < Me$_3$CNC ~ pMeO.C$_6$H$_4$.NC. N$_2$ then is a poor σ donor and/or good π acceptor.

The plot of centre shift versus quadrupole splitting enables a qualitative separation of σ and π. Except for N$_2$ and CO, a reasonable straight line correlation is observed (Fig. 6.3) with a *positive* slope. As ($\sigma + \pi$) increases (the centre shift decreases), ($\sigma - \pi$) increases (the quadrupole splitting becomes more negative). If both the centre shift and quadrupole splitting had comparable sensitivities to both σ and π effects, then a correlation would not be observed. Since CO and isocyanides are known to be better σ donors and π acceptors than nitriles from the centre shift values and general chemistry, the quadrupole splittings must be mainly sensitive to the σ donor power of the ligands. If the π term dominated the quadrupole splittings in this series, and π increased from left to right, then a line of opposite slope to that observed above would be obtained. N$_2$ and CO have substantially more positive quadrupole splittings than they should have to fall on the line in Fig. 6.3. This strongly indicates that these ligands are strong π acceptors

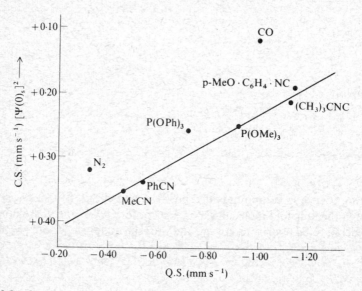

Fig. 6.3 Centre shift versus quadrupole splitting for *trans*-[FeHL(depe)$_2$]$^+$BPh$_4^-$ compounds (Tables 5.3 and 6.3).

relative to their neighbouring ligands. Thus, the centre shift and quadrupole splitting evidence taken together shows that N_2 is a comparable $(\sigma + \pi)$ ligand to the nitriles, but that it is an appreciably better π acceptor and poorer σ donor than the nitriles. Similarly, CO is a better π acceptor but comparable σ donor to the isocyanides. This evidence suggests that relative to these other neutral ligands, N_2 is a very weak σ donor, but a moderate π acceptor.

A somewhat similar plot of centre shift versus quadrupole splitting for other Fe^{II} low spin compounds of general formula trans-$Fe(niox)_2YZ$ (Fig. 6.4) lends support to the above arguments. The signs of the quadrupole

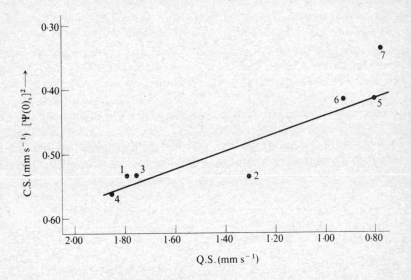

Fig. 6.4 Centre shift versus quadrupole splitting for $Fe(niox)_2YZ$ compounds [B. W. Dale, R. J. P. Williams, P. R. Edwards, and C. E. Johnson, Trans. Farad. Soc. **64**, 620 (1968); ibid., **64**, 3011 (1968).] 1. $Fe(niox)_2(Py)_2$; 2. $Fe(niox)_2(Im)_2$; 3. $Fe(niox)_2(NH_3)_2$; 4. $Fe(niox)_2(but)_2$; 5. $K_2[Fe(niox)_2(CN)_2]$; 6. $K[Fe(niox)_2Im.CN]$; 7. $Fe(niox)_2Im.CO$. niox = 1,2-cyclohexanedione dioxime; Py = pyridine; Im = imidazole; but = n-butylamine.

splitting for these compounds are positive, and the Z EFG axis is taken through the four-fold molecular axis. Except for CO, Y and Z are normally considered to be poor π acceptors, and thus the quadrupole splitting should become less positive (or more negative) as the σ donor powers of Y and Z increase. As expected then, the stronger σ donor, CN^-, gives a more negative quadrupole splitting than a moderate σ donor such as NH_3 or pyridine. The compound (#7) containing CO once again has a much more positive value than expected for a linear Q.S.–C.S. correlation. This indicates the

Mössbauer spectroscopy

importance of π acceptance in determining quadrupole splitting values for such strong π acceptors.

A similar treatment can be used to rationalize the quadrupole splittings for the $[Ru(NH_3)_6 X^{-n}]^{(2-n)+}$ and $[Ru(X)_5 NO]^{2-}$ compounds (Table 6.4). Again the Z EFG axis is taken through the four-fold axes for the above species.

Table 6.4 Q.S. values for Ru^{II} compounds (mm s^{-1} at 4·2 K)

Compound	Q.S.	Compound	Q.S.
$[Ru(NH_3)_6]Cl_2$	small	$K_2[Ru(CN)_5NO]$	0·42†
$[Ru(NH_3)_5(CH_3CN)](ClO_4)_2$	small	$Rb_2[Ru(NCS)_5NO]$	0·24
$[Ru(NH_3)_5N_2]X_2$	0·24†	$Rb_2[RuCl_5NO]$	0·24
$[Ru(NH_3)_5SO_2]Cl_2$	0·30	$Cs_2[RuBr_5NO]$	0·08
$[Ru(NH_3)_5CO]Br_2$	small	$K_4[Ru(CN)_5NO_2]2H_2O$	0·28
$[Ru(NH_3)_5NO]Cl_3.H_2O$	0·36†		

† Average values from references below.
W. Potzel, F. E. Wagner, U. Zahn, R. L. Mössbauer and J. Danon, *Z. Physik*, **240**, 306 (1970); C. A. Clausen, R. A. Prados and M. L. Good, *J. Am. Chem. Soc.*, **92**, 7483 (1970); R. A. Prados, Ph.D. Thesis, Louisiana State University, New Orleans (1971); R. Greatrex, N. N. Greenwood and P. Kaspi, *J. Chem. Soc. (A)*, 1873 (1971).

We use the bonding information from the Fe^{II} low spin compounds in an attempt to rationalize the sign and relative magnitudes of the quadrupole splitting values.

Considering the Ru ammine complexes, we would expect a zero quadrupole splitting for $[Ru(NH_3)_6]Cl_2$ just as the quadrupole splitting in the cubic $Fe(CN)_6^{4-}$ was expected to be zero. When X is substituted for NH_3 along the Z EFG axis, a negative q (and quadrupole splitting) will be obtained if X is a better σ donor than NH_3 and/or if X is a poorer π acceptor than NH_3, i.e., the quadrupole splitting becomes more negative as $(\sigma - \pi)$ of X increases. For $X = NO^+$ we expect a large positive quadrupole splitting because of the strong π acceptor strength of NO^+. The comparatively large quadrupole splitting for the N_2 compound is perhaps unexpected at first glance, but N_2 is most likely a poorer σ donor and better π acceptor than NH_3. Both σ and π effects contribute to a positive quadrupole splitting. In contrast, CO is probably a better σ donor and better π acceptor than NH_3, and the σ and π terms cancel out. A small unresolvable quadrupole splitting results.

For the $[RuX_5NO]$ series, the quadrupole splittings are most probably positive because of the very strong π acceptor properties of NO^+ relative to the X ligands. The quadrupole splitting becomes more positive in the order $X = Br^- < Cl^- < NCS^- < NH_3 < CN^-$ (Table 6.4) and a good correlation of centre shift versus quadrupole splitting is observed (Fig. 6.5).

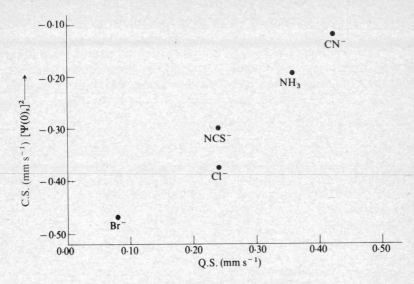

Fig. 6.5 Centre shift versus quadrupole splitting for [RuX$_5$NO]$^=$ compounds (from R. Greatrex *et al. J. Chem. Soc. (A)*, 1873 (1971).

Again $[\Psi(0)_s]^2$ increases as the σ donor capacity of X increases, and as for the Fe^{II} low spin compounds, the differences in σ donor abilities (rather than π acceptor properties) of X appear to dominate the differences in quadrupole splittings.

In passing, it is interesting to note that the ratio of quadrupole splittings for the analogous NO$^+$ and NO$_2^-$ pentacyanide complexes of Fe^{II} and Ru^{II} are very similar. For example, the ratio of Q.S.$_{NO^+}$/Q.S.$_{NO_2^-}$ for Fe^{II} is $1\cdot 73/0\cdot 89 = 1\cdot 94$, while the corresponding ratio for the ruthenium complexes is $0\cdot 42/0\cdot 28 = 1\cdot 50$. The latter value has a large uncertainty because of the large uncertainties in the small Ru quadrupole splittings. However, the similar ratios indicate that the d orbital imbalance, and therefore the bonding, in corresponding Fe^{II} and Ru^{II} compounds is similar.

6.2 Partial quadrupole splittings: the 2 : –1 *trans:cis* ratio

Partial quadrupole splittings (p.q.s.) and the additivity model have greatly facilitated the interpretation of quadrupole splittings for Fe^{II} low spin, Fe^{-II}, and Sn^{IV} compounds, and this approach should be very useful in rationalizing quadrupole splittings for other *N.Q.R.* and Mössbauer species such as Fe^{III} high spin, Ir^{III}, Ru^{II}, W^0, W^{VI}, Co^{III}, Te^{VI}, and Mn^{I}. The additivity model is expected to apply for compounds of transition metal ions whose t_{2g} and/or e_g subshells are filled or half filled, e.g., Fe$^{II}(t_{2g}^6)$; or for compounds of main group ions whose s and p shells are empty e.g., Sn^{IV}, Sb^V, Te^{VI},

($4d^{10}$). These ions all give cubic or spherical electronic symmetry about the Mössbauer nucleus.

For the above ions, $q_{C.F.} = 0$, and quadrupole splittings will be purely a function of the nature and distribution of the ligand bonds. The large range of quadrupole splittings observed are largely due to both p and/or d orbital imbalances (eqs. 2.14 and 2.15) due to different bonding properties of ligands, and the different structural types. For example, as seen in chapter 2 (eqs 2.10b and 2.11b) a *trans*-MA_2B_4 isomer should have twice the quadrupole splitting of a *cis*-MA_2B_4 isomer. Similarly, four-, five- and six-coordinate Sn compounds have characteristic ranges of quadrupole splittings, and the partial quadrupole splittings are of considerable use in rationalizing these different values. The partial quadrupole splittings can then be used in estimating bonding properties of ligands.

We assume that a ligand L on the Z EFG axis always gives the same contribution to the quadrupole splitting for a given central metal atom M in one coordination number. This contribution (given by $[L]$ values in eq. 2.10 and 2.11 and Table 2.3) is termed the partial field gradient (p.f.g.). The partial quadrupole splitting (p.q.s.) is given in terms of the partial field gradient by:

$$\text{p.q.s.} = \tfrac{1}{2} e^2 |Q|[L] \tag{6.8}$$

and partial quadrupole splittings are given in mm s^{-1}. The constant partial quadrupole splitting implies that the $M-L$ bond distance should be approximately a constant from compound to compound and/or that the quadrupole splitting is insensitive to small changes in bond lengths or bonding properties of L. We also assume that the complexes are of the highest possible symmetry: i.e. that a *trans*-MA_2B_4 compound has D_{4h} symmetry.

As for the iodine compounds discussed in section 6.1, it is assumed that the quadrupole splitting arises solely from intramolecular bonding; i.e. that the contribution to the quadrupole splitting from parts of the lattice other than nearest neighbours can be neglected. This assumption is most likely to break down for cationic and anionic compounds where a significant q_{lattice} contribution from the anion or cation respectively may arise. The point charge calculations (eqs. 2.10 and 2.11 and Table 2.3) for *trans*- and *cis*-MA_2B_4 isomers indicated that the ratio of *trans*:*cis* quadrupole splittings should be $2:-1$ if our above assumptions hold. Thus:

$$V_{zz}(\text{trans}) = (4[A] - 4[B])e \tag{2.10b}$$

$$V_{zz}(\text{cis}) = (-2[A] + 2[B])e \tag{2.11b}$$

Examination of the quadrupole splittings in Table 6.5 shows that the 2:1 ratio does indeed hold for three sets of Fe^{II} isomers, and magnetic spectra of *trans*- and *cis*-FeCl$_2$(ArNC)$_4$ (Fig. 6.6) shows that the *trans*-isomer has a

Bonding and structure from quadrupole splittings

Table 6.5 Quadrupole splittings for Fe^{II} low spin *cis-trans* isomers (mm s^{-1} at 295 K)

Compound	Q.S.
trans-FeCl$_2$(ArNC)$_4$	+1.55
cis-FeCl$_2$(ArNC)$_4$	−0.78
trans-Fe(SnCl$_3$)$_2$(ArNC)$_4$	(+)1.05
cis-Fe(SnCl$_3$)$_2$(ArNC)$_4$	(−)0.50
trans-Fe(CN)$_2$(EtNC)$_4$	−0.60
cis-Fe(CN)$_2$(EtNC)$_4$	(+)0.30

ArNC = *p*-methoxyphenylisocyanide.
G. M. Bancroft, M. J. Mays and B. E. Prater, *J. Chem. Soc.* (*A*), 956 (1970); R. R. Berrett and B. W. Fitzsimmons, *J. Chem. Soc.* (*A*), 525 (1967).

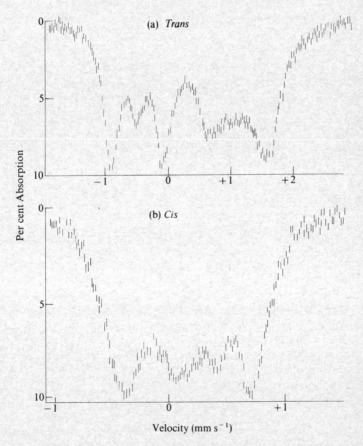

Fig. 6.6 Magnetic spectra of *cis* and *trans*-FeCl$_2$(ArNC)$_4$. [G. M. Bancroft, R. E. B. Garrod, A. G. Maddock, M. J. Mays, and B. E. Prater, *Chem. Comm.*, 201 (1970).]

positive quadrupole splitting, (the doublet lies to positive velocities), while the *cis*-isomer has a negative quadrupole splitting. The simple, but unrealistic, point charge approach (where Z_A and Z_B are "effective" charges) thus is successful in rationalizing the $2:-1$ ratio.

By fixing or calculating one partial quadrupole splitting, it is then possible to calculate a series of partial quadrupole splittings. Before doing this, it is worthwhile to examine a simple molecular orbital approach for rationalizing the $2:-1$ ratio. Other, more rigorous approaches are discussed in reference 6.2. The MO approach is chemically much more satisfying than the point charge model, and leads to the use of partial quadrupole splittings for obtaining valuable bonding information.

Consider a six-coordinate complex such as *trans*- or *cis*-MA_2B_4. As discussed earlier in chapter 5, bonding between ligand and metal can be represented by an overlap between d^2sp^3 hybrid metal orbitals with the σ orbitals of the ligands.

The octahedral hybrid orbitals directed along the ligand axes are:

$$\phi_1 = (1/\sqrt{6})s + (1/\sqrt{2})p_z + (1/\sqrt{3})d_{z^2} \tag{6.9a}$$

$$\phi_2 = (1/\sqrt{6})s - (1/\sqrt{2})p_z + (1/\sqrt{3})d_{z^2} \tag{6.9b}$$

$$\phi_3 = (1/\sqrt{6})s + (1/\sqrt{2})p_x - (1/\sqrt{12})d_{z^2} + (1/2)d_{x^2-y^2} \tag{6.9c}$$

$$\phi_4 = (1/\sqrt{6})s - (1/\sqrt{2})p_x - (1/\sqrt{12})d_{z^2} + (1/2)d_{x^2-y^2} \tag{6.9d}$$

$$\phi_5 = (1/\sqrt{6})s + (1/\sqrt{2})p_y - (1/\sqrt{12})d_{z^2} - (1/2)d_{x^2-y^2} \tag{6.9e}$$

$$\phi_6 = (1/\sqrt{6})s - (1/\sqrt{2})p_y - (1/\sqrt{12})d_{z^2} - (1/2)d_{x^2-y^2} \tag{6.9f}$$

Using eqs (2.14) and (2.15), and neglecting π bonding, $q_{M.O.}$ becomes:

$$q_{M.O.} = K_p[-N_{p_z} + \tfrac{1}{2}(N_{p_x} + N_{p_y})] + K_d[-N_{d_{z^2}} + N_{d_{x^2-y^2}}]. \tag{6.10}$$

The s orbitals, being spherically symmetric, do not contribute to $q_{M.O.}$. For the *cis*- and *trans*-MA_2B_4 isomers, we determine relative populations ($\propto N$'s) of the above hybrid orbitals. The Z EFG axes for these two isomers are defined as in Fig. 2.5. The molecular orbitals between the metal and ligands (Fig. 5.2) are a combination of the above hybrids and ligand orbitals. The population of each p and d orbital within the above hybrids is given by the square of the coefficients (Table 6.6) (which express the relative amount of charge density in a given ligand direction) multiplied by the donor strength of the ligand (denoted σ_L) (Table 6.7) and summing over all hybrid orbitals. Using eq. (6.10), and adding up the total population of the p and d orbitals, we obtain:

$$q_{M.O.}(trans) \propto K_p[\sigma_B - \sigma_A] + K_d[2/3\sigma_B - 2/3\sigma_A] \tag{6.11}$$

$$q_{M.O.}(cis) \propto K_p[-1/2\sigma_B + 1/2\sigma_A] + K_d[-1/3\sigma_B + 1/3\sigma_A]. \tag{6.12}$$

Table 6.6 Relative charge densities in p and d orbitals of the 6 octahedral σ bonding hybrids

Ligand	p_x	p_y	p_z	$d_{x^2-y^2}$	d_{z^2}
L_Z	0	0	$\frac{1}{2}$	0	$\frac{1}{3}$
L_Z	0	0	$\frac{1}{2}$	0	$\frac{1}{3}$
L_X	$\frac{1}{2}$	0	0	$\frac{1}{4}$	$\frac{1}{12}$
L_X	$\frac{1}{2}$	0	0	$\frac{1}{4}$	$\frac{1}{12}$
L_Y	0	$\frac{1}{2}$	0	$\frac{1}{4}$	$\frac{1}{12}$
L_Y	0	$\frac{1}{2}$	0	$\frac{1}{4}$	$\frac{1}{12}$

Table 6.7 Relative changes in orbital populations for *trans*- and *cis*-MA_2B_4

trans-MA_2B_4

Ligand (Figure 2.5)	p_x	p_y	p_z	$d_{x^2-y^2}$	d_{z^2}
A_1	0	0	$\frac{1}{2}\sigma_A$	0	$\frac{1}{3}\sigma_A$
A_2	0	0	$\frac{1}{2}\sigma_A$	0	$\frac{1}{3}\sigma_A$
B_1	$\frac{1}{2}\sigma_B$	0	0	$\frac{1}{4}\sigma_B$	$\frac{1}{12}\sigma_B$
B_3	$\frac{1}{2}\sigma_B$	0	0	$\frac{1}{4}\sigma_B$	$\frac{1}{12}\sigma_B$
B_2	0	$\frac{1}{2}\sigma_B$	0	$\frac{1}{4}\sigma_B$	$\frac{1}{12}\sigma_B$
B_4	0	$\frac{1}{2}\sigma_B$	0	$\frac{1}{4}\sigma_B$	$\frac{1}{12}\sigma_B$
Total	σ_B	σ_B	σ_A	σ_B	$\frac{2}{3}\sigma_A + \frac{1}{3}\sigma_B$

cis-MA_2B_4

Ligand (Figure 2.5)	p_x	p_y	p_z	$d_{x^2-y^2}$	d_{z^2}
B_1	0	0	$\frac{1}{2}\sigma_B$	0	$\frac{1}{3}\sigma_B$
B_2	0	0	$\frac{1}{2}\sigma_B$	0	$\frac{1}{3}\sigma_B$
A_1	$\frac{1}{2}\sigma_A$	0	0	$\frac{1}{4}\sigma_A$	$\frac{1}{12}\sigma_A$
B_3	$\frac{1}{2}\sigma_B$	0	0	$\frac{1}{4}\sigma_B$	$\frac{1}{12}\sigma_B$
A_2	0	$\frac{1}{2}\sigma_A$	0	$\frac{1}{4}\sigma_A$	$\frac{1}{12}\sigma_A$
B_4	0	$\frac{1}{2}\sigma_B$	0	$\frac{1}{4}\sigma_B$	$\frac{1}{12}\sigma_B$
Total	$\frac{1}{2}(\sigma_A+\sigma_B)$	$\frac{1}{2}(\sigma_A+\sigma_B)$	σ_B	$\frac{1}{2}(\sigma_A+\sigma_B)$	$\frac{1}{6}\sigma_A + \frac{5}{6}\sigma_B$

(G. M. Bancroft and R. H. Platt, *Adv. Inorg. Radiochem.*, **15**, 59, (1972)).

For example, the population of the $d_{x^2-y^2}$ orbital in the *trans* isomer will be proportional to $4(1/4\sigma_B) = \sigma_B$, while the population of the d_{z^2} orbital will be proportional to $2(1/3\sigma_A) + 4(1/12\sigma_B) = 2/3\sigma_A + 1/3\sigma_B$. $K_d[-N_{d_{z^2}} + N_{d_{x^2-y^2}}]$ becomes $\propto K_d[-2/3\sigma_A - 1/3\sigma_B + \sigma_B] = K_d[2/3\sigma_B - 2/3\sigma_A]$, which is given on the right hand side of eq. (6.11). The other terms in eqs (6.11) and (6.12) are obtained in a similar way.

The above two equations show that the 2:−1 ratio is expected on the basis of σ bonding to d^2sp^3 hybrids, or just to p and/or d orbitals. The magnitude of the quadrupole splitting is directly related to the differences in σ donor characteristics of the two ligands.

Similarly if π back-donation from the t_{2g} d orbitals is considered, $q_{M.O.}$ is given by:

$$q_{M.O.} = K_d[N_{d_{xy}} - \tfrac{1}{2}(N_{d_{xz}} + N_{d_{yz}})] \tag{6.13}$$

We take π_A and π_B to represent the π acceptor ability of A and B respectively. The three t_{2g} orbitals have equal charge distributions along the six bonding directions, and as discussed in chapter 2 (Fig. 2.7) each ligand π bonds to two of the t_{2g} d orbitals. In contrast to σ bonding, where the metal orbital populations increased as σ_L increased, the t_{2g} orbital populations will decrease as π_L increases. Adding up the orbital populations as for σ bonding and substituting into equation 6.13, we obtain:

$$q_{M.O.}(trans) \propto K_d(-2\pi_B + 2\pi_A) \tag{6.14}$$

$$q_{M.O.}(cis) \propto K_d(+\pi_B - \pi_A) \tag{6.15}$$

We can then combine eqs (6.11) and (6.14), and (6.12) and (6.15) to obtain:

$$q_{M.O.}(trans) \propto K_p[\sigma_B - \sigma_A] + K_d[2/3\sigma_B - 2/3\sigma_A - 2\pi_B + 2\pi_A] \tag{6.16}$$

$$q_{M.O.}(cis) \propto K_p[-1/2\sigma_B + 1/2\sigma_A] + K_d[-1/3\sigma_B + 1/3\sigma_A + \pi_B - \pi_A] \tag{6.17}$$

and

$$q_{M.O.}(trans) : q_{M.O.}(cis) = 2 : -1$$

It is apparent then from eqs (2.10), (2.11), (6.16) and (6.17), that the $2:-1$ ratio can be rationalized both on the basis of a point charge model or a molecular orbital model. Calculations on other geometric isomers and other coordination numbers show that the point charge and molecular orbital calculations usually give the same or equivalent predictions (refs. 6.1 and 6.2).

In forthcoming sections, the point charge model is used for convenience for deriving partial quadrupole splittings and predicting quadrupole splittings. The derived partial quadrupole splittings can then be used as an indication of the relative σ donor and π acceptor properties of ligands. For example, a ligand which concentrates negative charge along the Z axis should have a relatively negative partial quadrupole splitting, and we can write:

$$(p.q.s.)_L \propto -(pc)_L - (\sigma_L - \pi_L) \tag{6.18}$$

where: $(pc)_L$ is the point charge contribution to the partial quadrupole splitting. Thus a negative point charge or a strong σ donor on the Z EFG axis concentrate negative charge along the Z EFG axis, and give negative contributions to the partial quadrupole splitting. A π accepting ligand such as

Bonding and structure from quadrupole splittings

CO on the Z EFG axis removes electron density from the Z axis, and gives a positive contribution to the partial quadrupole splitting. After partial quadrupole splittings for ligands bound to Sn and Fe are derived in the following paragraphs, eq. (6.18) will be used to discuss bonding properties of ligands in section four of this chapter.

6.3 Derivation of partial quadrupole splittings: rationalization of signs and magnitudes of quadrupole splittings

The results in Table 6.5 showed that the theoretically predicted $2:-1$ *trans*:*cis* ratio is observed experimentally, showing that our initial assumptions regarding partial quadrupole splittings are reasonable. However, before using derived partial quadrupole splittings for structural predictions, we want to check theoretically predicted and observed values more closely. In the next few paragraphs, partial quadrupole splittings are derived for ligands bonded to Fe^{II} and Sn^{IV} compounds, and calculated and observed quadrupole splittings are compared for 'cross-check' compounds containing the same ligands in different combinations. Agreement between predicted and observed values is generally very good. The Fe^{II} partial quadrupole splittings are then used for 'determining' signs of the quadrupole splittings for Co^{III}, Ir^{III} and Ru^{II} compounds.

Fe^{II} *low spin compounds*. To set up a scale of partial quadrupole splitting values, we first must define one partial quadrupole splitting which we term a reference value. Equations (2.10) and (2.11) show that the quadrupole splittings for *cis-trans* isomers are proportional to *differences* in partial field gradients, so that a reference value cannot be obtained from one quadrupole splitting. The partial quadrupole splitting for Cl^- was initially calculated to be -0.30 mm s^{-1}, and we use this as the reference value. This value is not expected to be accurate, but it seems to be better than one obtained by arbitrarily assigning a reference value.†

Taking the above reference value for $(p.q.s.)_{Cl}$, the partial quadrupole splittings in Table 6.9 have been derived using the quadrupole splittings on the left hand side of Table 6.8, and using equations 2.10, 2.11, 6.8 and the equations for *trans*-$FeACB_4$ compounds and $[FeAB_5]^+$ compounds (Table 6.13):

$$V_{zz}(ACB_4) = (2[A] + 2[C] - 4[B])e \qquad (6.19)$$

$$V_{zz}(AB_5) = (2[A] - 2[B])e \qquad (6.20)$$

† Since the quadrupole splittings are always given by *differences* in partial quadrupole splittings, the reference value is not of great importance. *Relative* partial quadrupole splittings and *predicted* quadrupole splittings—at least for four- and six-coordinate compounds—are *independent* of the choice of a reference value.

Mössbauer spectroscopy

Table 6.8 Fe(II) low spin room temperature quadrupole splittings (mm s^{-1})

Compound	Q.S.	Compound	Q.S. Observed	Q.S. Calculated
1. trans-FeCl$_2$(ArNC)$_4$	+1·55	17. cis-FeCl$_2$(ArNC)$_4$	−0.78	−0.78
2. trans-Fe(SnCl$_3$)$_2$(ArNC)$_4$	+1·05	18. [FeCl(ArNC)$_5$]ClO$_4$	0.73	(+)0·78
3. trans-FeCl$_2$(depe)$_2$	+1·29	19. cis-Fe(SnCl$_3$)$_2$(ArNC)$_4$	0.50	(−)0·52
4. trans-FeCl$_2$(dmpe)$_2$	+1·51	20. [Fe(SnCl$_3$)(ArNC)$_5$]ClO$_4$	0.32	(+)0·52
5. trans-FeBr$_2$(depe)$_2$	(+)1·37	21. trans-FeHCl(depe)$_2$	≤0.12	(−)0·20
6. trans-FeI$_2$(depe)$_2$	(+)1·33	22. trans-FeHI(depe)$_2$	≤0.19	(−)0·18
7. trans-FeCl$_2$(depb)$_2$	(+)1·13	23. trans-FeClSnCl$_3$(depe)$_2$	1·28	(+)1·02
8. trans-FeH$_2$(depb)$_2$	(−)1·84	24. trans-FeBr$_2$(depb)$_2$	1·22	(+)1·20
9. cis-FeBr$_2$(CO)$_4$	(−)0·28	25. cis-Fe(CN)$_2$(EtNC)$_4$	0·29	(+)0·30
10. trans-Fe(NCO)$_2$(depe)$_2$	(+)0·49	26. cis-H$_2$Fe(CO)$_4$	0·55	(+)1·22
11. trans-Fe(NCS)$_2$(depe)$_2$	(+)0·53			
12. trans-Fe(N$_3$)$_2$(depe)$_2$	(+)0·98			
13. trans-Fe(EtNC)$_4$(CN)$_2$	−0·60			
14. Na$_2$[Fe(CN)$_5$NO·2H$_2$O	+1·73			
15. Na$_3$[Fe(CN)$_5$NH$_3$]·H$_2$O	+0·67			
16. K$_3$[Fe(CN)$_5$H$_2$O]·7H$_2$O	+0·80			

G. M. Bancroft, M. J. Mays and B. E. Prater, *J. Chem. Soc.* (A), 956 (1970).

* Signs in brackets are assumed or calculated; those without brackets have been measured using the magnetic field technique.

Bonding and structure from quadrupole splittings

Table 6.9 Partial quadrupole splitting (p.q.s.) values for ligands in Fe^{II} low-spin compounds (mm s^{-1} at 295 K)

Ligand	p.q.s.	Ligand	p.q.s.
NO^+	+0.02	NCO^-	−0.50
Br^-	−0.28	NH_3	−0.51
I^-	−0.29	PPh_3	∼ −0.5
Cl^-	−0.30	depb/2	−0.58
N_3^-	−0.38	depe/2	−0.62
$SnCl_3^-$	−0.43	dmpe/2	−0.67
CO	−0.55	RNC	−0.69
H_2O	−0.44	CN^-	−0.84
NCS^-	−0.49	H^-	−1.04

G. M. Bancroft, R. E. B. Garrod and A. G. Maddock, *J. Chem. Soc.* (*A*), 3165 (1971); G. M. Bancroft and E. T. Libbey, unpublished results.

To derive a partial quadrupole splitting for ArNC, we use the quadrupole splitting for *trans*-FeCl$_2$(ArNC)$_4$ and write: +1.55 = 4(p.q.s.)$_{Cl}$ − 4(p.q.s.)$_{ArNC}$(mm s^{-1}). Substituting in (p.q.s.)$_{Cl}$ = −0.30 mm s^{-1}, we obtain (p.q.s.)$_{ArNC}$ = −0.69 mm s^{-1}. Similarly, the partial quadrupole splitting for $SnCl_3^-$ is calculated from the quadrupole splitting for *trans*-Fe(SnCl$_3$)$_2$(ArNC)$_4$ along with the above (p.q.s.)$_{ArNC}$ value, i.e. +1.05 mm s^{-1} = 4(p.q.s.)$_{SnCl_3^-}$ − 4(p.q.s.)$_{ArNC}$. Solving, we obtain (p.q.s.)$_{SnCl_3^-}$ = −0.43 mm s^{-1}. The remaining partial quadrupole splittings are derived in a similar fashion (Table 6.9). Then using these partial quadrupole splittings, calculated and observed quadrupole splittings are compared for compounds 17–26 on the right hand side of Table 6.8. The agreement between predicted and observed values is very good (within 0.08 mm s^{-1}) except for compounds 20, 23, and 26. The first discrepancy (compound 20) might be associated with the cationic compound, and a significant $q_{lattice}$ contribution from the anion. The latter discrepancy (compound 26) is almost certainly associated with the variable bonding properties (and resulting variable partial quadrupole splitting) of strong π accepting ligands such as CO. The partial quadrupole splitting for CO reported in Table 6.9 should be regarded as a preliminary working value and it has been suggested (ref. 6.2) that the additivity model might not hold for strong π acceptors. However, for the great majority of ligands, self-consistent partial quadrupole splittings can be derived, and agreement between predicted and observed quadrupole splittings for six-coordinate Fe^{II} compounds is usually good.

The partial quadrupole splittings are now very helpful for predicting the sign of the quadrupole splitting in situations where the sign is not known or

Mössbauer spectroscopy

difficult to determine. The signs of many of the quadrupole splittings in Table 6.3 were assigned on the basis of partial quadrupole splittings. The predicted quadrupole splitting for *trans*-[FeH(pMeO.C$_6$H$_4$.NC) (depe)$_2$]$^+$-BPh$_4^-$ using the eq. (6.19) and the partial quadrupole splittings in Table 6.9 is -0.98 mm s^{-1} compared to the observed value of -1.14 mm s^{-1}. With this good agreement, and since the partial quadrupole splittings for all neutral ligands should be in the observed range -0.30 mm s^{-1} to -0.70 mm s^{-1} (Table 6.9), the quadrupole splittings for *trans*-[FeHL(depe)$_2$]$^+$ species should be in the range -0.20 mm s^{-1} [(p.q.s.)$_L$ = -0.30 mm s^{-1}] to -1.00 mm s^{-1} [(p.q.s.)$_L$ = -0.70 mm s^{-1}], and this is just the range of quadrupole splittings observed. Thus it seems highly likely that the quadrupole splittings for [FeHL(depe)$_2$]$^+$ compounds in Table 6.3 are negative. The sign of the quadrupole splitting for the N$_2$ complex would be difficult to obtain in any other way because of its small magnitude. If we derive a partial quadrupole splitting for N$_2$ from the N$_2$ complex, we obtain (p.q.s.)$_{N_2}$ = -0.36 mm s^{-1}, and this value is consistent with N$_2$ being a poorer σ donor and/or better π acceptor than NH$_3$ and MeCN—as mentioned in section one of this chapter.

Similarly, for the Fe(CN)$_5$X (Table 6.3) series, the partial quadrupole splittings for the ligands in Table 6.9 strongly suggest that all of these compounds have positive quadrupole splittings. For example, the predicted quadrupole splitting for [Fe(CN)$_5$(PPh$_3$)]$^{3-}$ is approximately [2(-0.5) $-$ 2(-0.84)] = $+0.7$ mm s^{-1}, in good agreement with the observed value of 0.62 mm s^{-1}. For the signs of these quadrupole splittings to be negative, the partial quadrupole splittings of PPh$_3$ and similar ligands would have to be as negative as the partial quadrupole splitting for H$^-$, and this seems highly unlikely from known chemical properties.

Co^{III}, Ir^{III} and Ru^{II} compounds. The Fe^{II} partial quadrupole splittings in Table 6.9 are also very useful for predicting signs and magnitudes of quadrupole splittings for Co^{III}, Ru^{II}, and Ir^{III} compounds. All of these ions, like Fe^{II}, have $(t_{2g})^6$ electronic ground states, and we might expect metal ligand bonding to be similar for all four ions. For example, we noted earlier in this chapter that the ratio of quadrupole splittings for the analogous Ru^{II} and Fe^{II} compounds, [M(CN)$_5$NO]$^=$ and [M(CN)$_5$NO$_2$]$^{4-}$, are similar as expected if the relative partial quadrupole splittings for NO$^+$, CN$^-$, and NO$_2^-$ are similar for both Ru^{II} and Fe^{II}. The signs of the Ru^{II} quadrupole splitting for these two compounds, like their Fe^{II} counterparts, are thus almost certainly positive. Similarly, all of the signs of the Ru^{II} quadrupole splittings in Table 6.4 can be predicted with some confidence using Fe^{II} partial quadrupole splittings. These signs cannot be determined experimentally by Mössbauer methods, because of the very small quadrupole splittings and comparatively poor spectra.

Table 6.10 ^{193}Ir quadrupole splittings for XYIrCl(CO)(PPh$_3$)$_2$ compounds (mm s^{-1} at 20 K)

X	Y	Q.S.
H	H	4.76
Cl	H	1.44
Cl	Cl	3.10

H. H. Wickman and W. E. Silverthorne, *Inorg. Chem.*, **10**, 2333 (1971).

Similarly, to take an example involving Ir^{III} in compounds of the type XYIrCl(CO)(PPh$_3$)$_2$ (Table 6.10), the partial quadrupole splittings in Table 6.9 probably could be used to predict the signs of these quadrupole splittings.

In a more quantitative treatment, the Fe^{II} partial quadrupole splittings can be used to predict the sign and magnitude of ^{59}Co^{III} quadrupole splittings; or if the quadrupole splittings for the corresponding Fe^{II} and Co^{III} compounds are known or can be calculated from partial quadrupole splittings, the ratio of quadrupole splittings can be used to obtain an estimate for $Q_{^{57}Fe}$. Cobalt does not have a Mössbauer nuclide, but quadrupole splittings can be obtained by nuclear quadrupole resonance (n.q.r.) or n.m.r. measurements. The sign of quadrupole splittings cannot be obtained directly by n.q.r. or n.m.r.

The quadrupole splittings for three Co^{III} compounds are given in Table 6.11. The quadrupole splittings for the corresponding Fe^{II} compounds are not known, because these compounds have not yet been made, and Fe(NH$_3$)$_4$Cl$_2$ and [Fe(NH$_3$)$_5$Cl]$^+$ would be high spin anyway. However, the quadrupole splitting for the three hypothetical Fe^{II} species can be estimated using

Table 6.11 Quadrupole splittings for Co^{III} and hypothetical Fe^{II} compounds

Compound	$\frac{1}{2}e^2qQ$ (mm s^{-1})	e^2qQ/h (Mc s^{-1})*
1. [Co(NH$_3$)$_5$CN](ClO$_4$)$_2 \cdot \frac{1}{2}$H$_2$O		(−)48.8
2. *trans*-[Co(NH$_3$)$_4$Cl$_2$]Cl		(+)59.23
3. [Co(NH$_3$)$_5$Cl]Cl$_2$		(+)31.74
4. '*trans*-Fe(NH$_3$)$_4$Cl$_2$'	+0.84	+19.5
5. '[Fe(NH$_3$)$_5$Cl]$^+$'	+0.42	+9.8
6. '[Fe(NH$_3$)$_5$CN]$^+$'	−0.67	−15.6

* For ^{57}Fe, 1 mm s^{-1} = 11.61 Mc s^{-1}.
G. M. Bancroft, *Chem. Phys. Lett.*, **10**, 449 (1971).

the partial quadrupole splittings for $Cl^-(-0.30\text{ mm s}^{-1})$, $NH_3(-0.51\text{ mm s}^{-1})$ and $CN^-(-0.84\text{ mm s}^{-1})$ from Table 6.9. Trans-$Fe(NH_3)_4Cl_2$ should have a quadrupole splitting of $[4(-0.30) - 4(-0.51)] = +0.84$ mm s^{-1}, and the other values are given in Table 6.11. Since Fe^{II} and Co^{III} are isoelectronic, it is reasonable to suggest that the bonding in corresponding Fe^{II} and Co^{III} compounds (e.g., trans-$[Co(NH_3)_4Cl_2]^+$ and 'trans-$Fe(NH_3)_4Cl_2$') is very similar, with the orbital populations in the two compounds differing by no more than ten per cent. Therefore the Co^{III} compounds should have the same sign of q and e^2qQ as the corresponding Fe^{II} compounds. (Table 6.11).

A more quantitative evaluation of the above results confirms the above assignments of sign and enables a semi-empirical calculation of the ^{57}Fe quadrupole moment. We take $Q_{^{59}Co} = 0.404b$, $eq_{zz} = -9.53 \times 10^{15}$ e.s.u. cm^{-3} for one Co d electron and $eq_{zz} = -8.08 \times 10^{15}$ e.s.u. cm^{-3} for one Fe d electron, and neglect the small contributions to e^2qQ from $4p$ orbital occupation. Assuming the bonding to be identical in trans-$Fe(NH_3)_4Cl_2$(1) and trans-$[Co(NH_3)_4Cl_2]^+$(2), we can write:

$$\frac{(e^2qQ)_2}{e^2q_{3d\text{ Co}}\,Q_{^{59}Co}} = \frac{(e^2qQ)_1}{e^2q_{3d\text{ Fe}}\,Q_{^{57}Fe}} \tag{6.21}$$

$$Q_{^{57}Fe} = \left[\frac{(e^2qQ)_1 q_{3d\text{ Co}}}{(e^2qQ)_2 q_{3d\text{ Fe}}}\right] Q_{^{59}Co} \tag{6.22}$$

Substituting the values of e^2qQ for both compounds, $Q_{^{59}Co}$, and q_{zz} for Fe and Co d electrons, we obtain $Q_{^{57}Fe} = 0.16b(\pm 0.03b)$. Slightly smaller results (0.15b) are obtained from the other two sets of compounds, and these values are in good agreement with that calculated ($Q_{^{57}Fe} = 0.175b$) using a similar method and the known e^2qQ values in $Fe(C_5H_5)_2$ and $Co(C_5H_5)_2^+$, but slightly smaller than the recognized value for $Q_{^{57}Fe}$ of about 0.20 barns.

From these results, it is apparent that Fe compounds have e^2qQ values about one-third those of the corresponding Co compounds. It should be possible now to use eqs (6.21) or (6.22) to calculate approximate e^2qQ values for either Co or Fe compounds, if the e^2qQ value for the corresponding Co or Fe compound is known, or can be predicted from partial quadrupole splittings.

For Ru^{II} and Ir^{III} compounds, equations analogous to (6.21) and (6.22) could be used for predicting Ru^{II} and Ir^{III} quadrupole splittings, but since it is not realistic to assume that bonding in Fe^{II} and the corresponding Ir^{III} and Ru^{II} compounds are identical, it would seem better to use the Fe^{II} partial quadrupole splittings to estimate relative magnitudes of the quadrupole splittings and their sign for Ru^{II} and Ir^{III} compounds.

Bonding and structure from quadrupole splittings

Sn^{IV} *compounds.* Self-consistent partial quadrupole splittings can also be derived for Sn^{IV} compounds. Initially, it was assumed by Parish and Platt that the same partial quadrupole splitting could be used for a ligand in four-, five- and six-coordinate Sn^{IV} compounds. These partial quadrupole splittings were extremely useful in rationalizing Sn^{IV} quadrupole splittings in general, but it soon became apparent that separate partial quadrupole splittings would have to be used for different coordination numbers. However, before looking at the most recent treatments of Sn^{IV} partial quadrupole splittings, it is important to examine the initial treatment. Table 6.12 gives the initial

Table 6.12 Quadrupole splittings, V_{zz} and initial p.q.s. values for ligands bonded to Sn^{IV} (mm s^{-1} at 80 K)

Structural type	Example	Q.S.*	V_{zz}/e	p.q.s. *Obtained* mm s^{-1}
Trigonal bipyramidal	$SnCl_5$	(−)0.63	[Cl]	Cl = +0.63
Trigonal bipyramidal (Cl axial)	R_3SnCl_2[a]	−3.46[b]	4[Cl] − 3[R]	R = −0.31
Trigonal bipyramidal	R_3SnF_2	−3.86[b]	4[F] − 3[R]	F = 0.73
Trigonal bipyramidal	R_3SnBr_2	−3.39[b]	4[Br] − 3[R]	Br = 0.60
Trigonal bipyramidal	Ph_3SnF_2[a]	−3.58[b]	4[F] − 3[Ph]	Ph = −0.20
Tetrahedral	Me_3SnCF_3	−1.38	2[CF_3] − 2[Me]	CF_3 = 0.38
Tetrahedral	$Me_2Sn(C_6F_5)_2$	1.49	2[C_6F_5] − 2[Me](1.15)	C_6F_5 = 0.32

* Remember that the quadrupole moment for ^{119}Sn is negative.
[a] R = Me, Et; Ph = C_6H_5. [b] These values are averages from [R_3SnX_2]$^-$ compounds and associated R_3SnX compounds.
R. V. Parish and R. H. Platt, *Inorg. Chim. Acta,* **4**, 65 (1970).

partial quadrupole splittings and the data used in deriving them. The point charge expressions given in Table 2.2 are used, keeping in mind that $Z_L/r_L^3 = [L]$, the partial field gradient of ligand L. V_{zz}/e for $SnCl_5^-$ is just given by [Cl] (assuming that axial and equatorial chlorines have the same partial field gradient values), and Q.S. = $\frac{1}{2}e^2Q[Cl]$ for $SnCl_5^-$. Since Q is negative for ^{119}Sn and p.q.s. = $\frac{1}{2}e^2|Q|[L]$ (eq. 6.8), then for ^{119}Sn, p.q.s. = $-\frac{1}{2}e^2Q[L]$, and (p.q.s.)$_{Cl}$ = −Q.S.$_{SnCl_5^-}$ = +0.63 mm s^{-1}. For ^{119}Sn, this value is a convenient reference value for deriving other partial quadrupole splittings. Then to derive (p.q.s.)$_R$, we take the negative of the quadrupole splitting for R_3SnCl_2 and the expression for V_{zz}/e calculated from the point charge expressions, and write: +3.46 = 4(p.q.s.)$_{Cl}$ − 3(p.q.s.)$_R$. Substituting the value calculated above for (p.q.s.)$_{Cl}$, we find that (p.q.s.)$_R$ = −0.31 mm s^{-1}.

Mössbauer spectroscopy

Similarly the partial quadrupole splittings for F, Br, Ph, CF_3, and C_6F_5 in Table 6.12 are calculated. The point charge expressions for the tetrahedral species are given along with many others of interest in Table 6.13. Where possible, the Z EFG axis is defined. In most cases, the Z EFG axis corresponds to the molecular axis of highest symmetry. Note that for five-coordinate species in Table 6.13, we use different partial field gradient values

Table 6.13 Point charge model expressions for the components of the EFG tensor for some common structures†

	Structure	Components of the EFG
1. MB_4	Z⋮B–M(B)(B)(B)	$V_{ZZ} = V_{XX} = V_{YY} = 0$
2. MAB_3	Z–A–M(B)(B)(B)	$V_{ZZ} = \{2[A] - 2[B]\}e$ $V_{YY} = \{-[B] + [A]\}e$ $V_{XX} = \{-[B] + [A]\}e$ $\eta = 0$
3. MA_2B_2[a]	A–M(A)(B)(B)	$V_{ZZ} = \{2[A] - 2[B]\}e$ $V_{YY} = \{2[B] - 2[A]\}e$ $V_{XX} = 0$ $\eta = 1$
4. MB_5	Z⋮B–M(B)(B)(B)–B⋯X	$V_{ZZ} = \{4[B]^{tba} - 3[B]^{tbe}\}e$ $V_{YY} = \{\frac{3}{2}[B]^{tbe} - 2[B]^{tba}\}e$ $V_{XX} = \{\frac{3}{2}[B]^{tbe} - 2[B]^{tba}\}e$ $\eta = 0$
5. MA_2B_3	Z⋮A–M(B)(B)(A)–B⋯X	$V_{ZZ} = \{4[A]^{tba} - 3[B]^{tbe}\}e$ $V_{YY} = \{-2[A]^{tba} + \frac{3}{2}[A]^{tbe}\}e$ $V_{XX} = \{-2[A]^{tba} + \frac{3}{2}[A]^{tbe}\}e$ $\eta = 0$

Table 6.13 (continued)

	Structure	Components of the EFG
6. MB_6	octahedral MB_6	$V_{ZZ} = V_{XX} = V_{YY} = 0$
7. trans-MA_2B_4	trans-MA_2B_4 structure	$V_{ZZ} = \{4[A] - 4[B]\}e$ $V_{YY} = \{2[B] - 2[A]\}e$ $V_{XX} = \{2[B] - 2[A]\}e$ $\eta = 0$
8. cis-MA_2B_4	cis-MA_2B_4 structure	$V_{ZZ} = \{2[B] - 2[A]\}e$ $V_{YY} = \{[A] - [B]\}e$ $V_{XX} = \{[A] - [B]\}e$ $\eta = 0$
9. MAB_5	MAB_5 structure	$V_{ZZ} = \{2[A] - 2[B]\}e$ $V_{YY} = \{[B] - [A]\}e$ $V_{XX} = \{[B] - [A]\}e$ $\eta = 0$
10. $MACB_4$	$MACB_4$ structure	$V_{ZZ} = \{2[A] + 2[C] - 4[B]\}e$ $V_{YY} = \{2[B] - [A] - [C]\}e$ $V_{ZZ} = \{2[B] - [A] - [C]\}e$ $\eta = 0$
11. MAB_3C_2	MAB_3C_2 structure	$V_{11} = \{2[A] - 2[C]\}e$ $V_{22} = \{4[C] - 3[B] - [A]\}e$ $V_{33} = \{3[B] - 2[C] - [A]\}e$ $\eta \neq 0$ (axes depend on relative magnitudes of $[A]$, $[B]$ and $[C]$

Mössbauer spectroscopy

Table 6.13 (continued)

Structure	Components of the EFG

12. $MA_2B_2C_2$ (trans A)

$V_{ZZ} = \{4[A] - 2[B] - 2[C]\}e$
$V_{YY} = \{[C] + [B] - 2[A]\}e$
$V_{XX} = \{[C] + [B] - 2[A]\}e$
$\eta = 0$

13. $MA_2B_2C_2$ (all cis)

$V_{11} = \{[B] + [C] - 2[A]\}e$
$V_{22} = \{[C] + [A] - 2[B]\}e$
$V_{33} = \{[B] + [A] - 2[C]\}e$
$\eta \neq 0$
axes depend on the relative magnitudes of $[A]$, $[B]$ and $[C]$

14. $MABC_2$[b]

$V_{zz} = \{2[A] - \frac{2}{3}([B] + 2[C])\}e$
$V_{yy} = \{-[A] - [B] + 2[C]\}e$
$V_{xx} = \{-[A] + \frac{5}{3}[B] - \frac{2}{3}[C]\}e$
$V_{xz} = V_{zx} = \{\frac{\sqrt{2}}{3}(-2[B] + 2[C])\}e$
$V_{xy} = V_{yx} = V_{yz} = V_{zy} = 0.$
$\eta \neq 0$

† For a more complete list, see ref. 6.1.
[a] The X axis coincides with the C_2 symmetry axis, and the Y and Z axes lie in the symmetry planes. [b] The Y axis is perpendicular to the symmetry plane, while the X and Z axes lie in the plane. The orientation of the X and Z axes depend upon the relative magnitudes of $[A]$, $[B]$, and $[C]$ and the tensor must be diagonalized separately for each case considered.

for axial and equatorial ligands (denoted tba and tbe respectively), in contrast to the assumption of equal partial field gradient values given above.

Using these preliminary partial quadrupole splittings (Table 6.12), the main structural trends in Sn^{IV} quadrupole splittings can be readily rationalized: i.e., that the magnitude of the quadrupole splitting generally decreases in the order: trans-octahedral > five-coordinate trigonal bipyramidal > cis-octahedral ~ four-coordinate ~ $RSnX_5^=$. The quadrupole splittings in Table 6.14 show this general trend: trans-R_2SnX_4 > R_3SnX_2 > cis-$R_2Sn(oxin)_2$ ~ $RSnX_3$ or R_3SnX ~ $RSnCl_5$, and the calculated quadrupole splittings on the right hand side of Table 6.14 using the partial quadrupole splittings in Table 6.12 generally give fair to good agreement with the observed results. For the last two compounds, Ph_3SnCl and Ph_3SnBr, the agreement between predicted and observed values is not satisfactory.

Table 6.14 Predicted and observed quadrupole splittings (mm s^{-1} at 80 K) using preliminary partial quadrupole splittings

	Q.S.	
Species	Observed	Calculated
trans-R$_2$SnF$_4^=$	+4·28	+4·16
trans-R$_2$SnCl$_4^=$	+4·21	+3·76
RSnCl$_5^=$	+1·90	+1·88
cis-R$_2$SnCl$_4^=$	not observed	−1·88
cis-R$_2$Sn(oxin)$_2$	+2·06	
R$_3$SnCl$_2$	−3·46	used in p.q.s. calculations
PhSnCl$_3$	1·84	+1·66
Ph$_3$SnCl(tet)	−2·54	−1·66
Ph$_3$SnBr(tet)	−2·50	−1·60

R. V. Parish and R. H. Platt, *Inorg. Chim. Acta*, **4**, 65 (1970). Averages of other values given in: G. M. Bancroft and R. H. Platt, *Adv. Inorg. Radiochem.*, **15**, 59 (1972).

For more accurate calculations, it has become apparent that separate partial quadrupole splittings for ligands in different coordination numbers should be derived and separate partial quadrupole splittings for the non-equivalent axial and equatorial atoms in trigonal bipyramidal coordination should also be used. Recently, separate partial quadrupole splittings have been derived for four- and six-coordinate Sn^{IV} compounds (Table 6.15) taking (p.q.s.)$_{F,Cl,Br}$ = 0. This is not meant to be an accurate value, but the previous value for (p.q.s.)$_{Cl}$ cannot be used because V_{ZZ}/e(SnCl$_5^-$) = 4[Cl]$_{tba}$ − 3[Cl]$_{tbe}$, and it is not possible to define either [Cl]$_{tba}$ or [Cl]$_{tbe}$. Also, it is desirable to have a separate reference value for each coordination number. It should also be remembered that for four- and six-coordination, relative partial quadrupole splittings and calculated quadrupole splittings are independent of what we choose as a reference value.

To obtain the partial quadrupole splittings in Table 6.15, a very similar procedure was used to that outlined previously in Table 6.12 and text. All alkyl groups were assumed to have the same partial quadrupole splitting. For example an average quadrupole splitting of −2·74 mm s^{-1} was obtained for R$_3$SnX species. Taking (p.q.s.)$_X$ = 0, then −2·74 = +2(p.q.s.)$_R$ and (p.q.s.)$_R$ = −1·37 mm s^{-1}. Similarly, to obtain (p.q.s.)$_{Ph}$, the average quadrupole splitting for Ph$_3$SnX species was found to be −2·52 mm s^{-1} and (p.q.s.)$_{Ph}$ = −1·26 mm s^{-1}. In a similar way, all the other partial quadrupole splittings were calculated. Having calculated these values, we can now use the equations in Table 6.13 to calculate quadrupole splittings for Sn^{IV} compounds containing these ligands in different combinations

Table 6.15 p.q.s. Values for Sn^{IV} compounds (mm s^{-1} at 80 K)[a]

Tetrahedral structures		Octahedral structures	
Ligand	p.q.s.	Ligand	p.q.s.
R	−1·37	R	−1·03
Ph	−1·26	Ph	−0·95
p-F.C$_6$H$_4$	−1·12	$\frac{1}{2}$(edt)[b]	−0·56
Fe(CO)$_2$Cp[c]	−1·08	$\frac{1}{2}${S$_2$CN(CH$_2$)$_4$}	−0·29
o-CF$_3$.C$_6$H$_4$	−1·04		
Mn(CO)$_5$[c]	−0·97	$\frac{1}{2}${S$_2$CNEt$_2$}	−0·25
C$_6$Cl$_5$	−0·83		
C$_6$F$_5$	−0·70	$\frac{1}{2}${S$_2$CNPh$_2$}	−0·23
CF$_3$	−0·63		
HCO$_2$	−0·18	$\frac{1}{2}${S$_2$CH(CH$_2$Ph)$_2$}	−0·19
I	−0·17	$\frac{1}{2}$(dipyam)[b]	−0·17
MeCO$_2$	−0·15	I	−0·14
F, Cl, Br	0·00	C$_5$H$_5$N	−0·10
NCS	+0·21	$\frac{1}{2}$(bipy)[b]	−0·08
		C$_5$H$_5$NO	−0·05
		$\frac{1}{2}$(phen)[b]	−0·04
		F, Cl, Br	0·00
		dmso[b]	+0·01
		NCS	+0·07

[a] For a more complete list, see references 6.1 and 6.2. [b] edt = ethanediol; dipyam = di-(2-pyridylamine); bipy = 2,2'-bipyridyl; dmso = dimethylsulphoxide; phen = 1,10-phenanthroline. [c] G. M. Bancroft, K. D. Butler, B. Dale and A. T. Rake, *J. Chem. Soc.* (Dalton), 2025 (1972).

Table 6.16 Predicted and observed quadrupole splittings for 4-coordinate and 6-coordinate Sn^{IV} compounds (mm s^{-1} at 80 K)

	Tetrahedral Q.S.			Octahedral Q.S.	
Compound	Obs.	Calc.	Compound	Obs.	Calc.
Ph$_3$SnI	2·15	−2·18	Bu$_2$SnBr$_2$phen	3·94	+4·04
Ph$_2$SnI(CH$_2$)$_4$SnIPh$_2$	−2·37	−2·26	Me$_2$SnF$_2$	+4·38	+4·12
Me$_2$Sn(C$_6$F$_5$)$_2$	1·51	1·55*	Ph$_2$SnCl$_2$(dmso)$_2$	3·54	+3·82
Ph$_3$SnC$_6$Cl$_5$	0·84	−0·86	Ph$_2$Sn(NCS)$_2$phen	+2·35	−2·26
Ph$_2$Sn(C$_6$Cl$_5$)$_2$	1·14	0·99*	(edt)$_2$Sn(C$_5$H$_5$N)$_2$	1·85	−1·84
PhSn(C$_6$F$_5$)$_3$	0·92	+1·12	[Bu$_2$Sn(NCS)$_4$]$^=$	4·35	+4·40
Me$_2$Sn[Mn(CO)$_5$]$_2$	0·92	0·92*	Ph$_2$Sn{S$_2$CNPh$_2$}$_2$	1·69	−1·44
Me$_2$ClSnMn(CO)$_5$	−2·59	−2·60			
MeCl$_2$SnMn(CO)$_5$	+2·79	+2·62			

For a more complete list, see references 6.1 and 6.2, and G. M. Bancroft, K. D. Butler, B. Dale and A. T. Rake, *J. Chem. Soc.* (Dalton), 2025 (1972).
* $\eta = 1$, sign is indeterminate.

(Table 6.16). The general agreement between predicted and observed values is excellent, especially because we have assumed ideal geometry; and for many of these compounds, significant distortions from ideal geometry exist. One very notable discrepancy has been found for cis-$Ph_2Sn(NCS)_2$phen and other cis-octahedral compounds, where the predicted and observed signs of the quadrupole splitting do not agree. This discrepancy has been attributed to significant distortion from ideal geometry which can change the sign without altering the magnitude of the quadrupole splitting appreciably.

Three other very important conclusions were reached in ref. 6.2 which will be used in further discussion. First, $(p.q.s.)_L^{oct}$ should be about 0.67 $(p.q.s.)_L^{tet}$ and this prediction is borne out by the results in Table 6.15. For example $(p.q.s.)_R^{oct}/(p.q.s.)_R^{tet} = 1.03/1.37 = 0.75$ and $(p.q.s.)_{Ph}^{oct}/(p.q.s.)_{Ph}^{tet} = 0.95/1.26 = 0.75$. This agreement between theory and experiment strongly supports the derivation and use of partial quadrupole splittings. Secondly, although five-coordination quadrupole splittings have not yet been considered in detail due to the difficulties in assigning different partial quadrupole splittings to axial and equatorial ligands, the ratio of splittings Q.S. (R_3SnX_2)/Q.S. (R_3SnX) should be about 1.33. The quadrupole splittings for the five-coordinate species Me_3SnX (X = F, Cl, Br, I) and the four-coordinate species Neo_3SnX show an average ratio of 1.28 in good agreement again with the predicted value. This large five-coordinate quadrupole splitting (noted earlier in Table 6.12 and text) is very useful in discussing intermolecular association in the next section. Thirdly, it was shown that π bonding, involving donation of electrons to the Sn $5d$ orbitals, would have no observable effect on the quadrupole splittings. Thus the variations in quadrupole splitting and partial quadrupole splittings must be interpreted solely in terms of σ bonding. This point is important for discussion in the following section.

6.4 Structural and bonding predictions from partial quadrupole splittings

The previous discussion for Fe^{II} low spin and Sn^{IV} quadrupole splittings shows that partial quadrupole splittings can be readily derived, and that they are a very useful book-keeping device for rationalizing signs and magnitudes of quadrupole splittings. However, they are probably of more important use in elucidating structural types (e.g. four- or five-coordinate R_3SnX compounds) and bonding properties of ligands. These two very important uses are discussed in the following two parts of this section.

Structure. It is often very difficult to predict the coordination number of Sn^{IV} compounds from the chemical formula. For example, it is perhaps surprising that Me_2SnF_2 and Me_3SnF contain six- and five-coordinate Sn^{IV}

respectively, while Ph_3SnCl and Ph_2SnCl_2 are both four-coordinate. It has even been difficult sometimes to tell the coordination number of the Sn from X-ray diffraction. For example, for Ph_2SnCl_2 the closest intermolecular Sn...Cl contact in the structure is 3·77 Å and this is almost certainly a non-bonding distance. However, for Me_2SnCl_2, which has been described as containing six-coordinate Sn^{IV}, it has been pointed out (ref. 6.3) that the distortion of the four nearest ligands is only slightly larger than that in Ph_2SnCl_2, and the intermolecular Sn...Cl contact is 3·54 Å. It is apparent then that there may be quite a subtle difference between a non-bonding and bonding distance, and it is not clear what a minimal non-bonding Sn–Cl distance might be.

We have shown in the previous section that five- and six-coordinate compounds should give markedly larger quadrupole splittings than four-coordinate values, and thus Mössbauer should be very useful in distinguishing between an unassociated and associated structure. In addition, extensive studies on the temperature dependence of the recoil-free fraction have been and will be, extremely useful for describing intermolecular interactions (for example, see ref. 6.4).

Consideration of the data in Tables 6.17 and 6.18 illustrates the great use of Mössbauer quadrupole splittings for obtaining structural information in

Table 6.17 Quadrupole splittings for $Ph_3SnOCOR'$ compounds (mm s^{-1} at 80 K)

R'	Q.S.	R'	Q.S.
1. $(CH_2)_3CHMe_2$	3·36	6. $CHMe_2$	3·32
2. $(CH_2)_2CHMe_2$	3·38	7. $CH=CH_2$	3·41
3. $CH_2CHMeEt$	3·39	8. $CMe=CH_2$	2·26
4. CH_2CHMe_2	3·39	9. $CHEtBu$	2·26
5. $CHMePr$	3·34	10. CMe_3	2·40

B. F. E. Ford and J. R. Sams, *J. Organomet. Chem.* **21**, 345 (1970).

Sn^{IV} species. For example, the branched chain triphenyl tin carboxylates (Table 6.17) of general formula $R_3SnOCOR'$ are generally linear polymers in the solid state with the five-coordinate Sn atom being coordinated to two *trans* oxygen atoms and three in-plane R groups. However, the quadrupole splittings for these compounds (Table 6.17) fall into two distinct groups: those having quadrupole splittings of ~3·3 mm s^{-1} (compounds 1–7), and three compounds (8, 9, 10) having quadrupole splittings of about 2·3 mm s^{-1}. These values are very similar to typical five-coordinate and four-coordinate values respectively given previously. For example, in Table 6.14, five-coordinate R_3SnCl_2 and four-coordinate R_3SnCl have quadrupole splittings of 3·46 and 2·54 mm s^{-1} respectively. In addition, for the carboxylates, the ratio of the Q.S.$_{1-7}$:Q.S.$_{8-10}$ ~ 1·4—in agreement with the ratio of ~1·33 expected (vide supra and ref. 6.2). Thus the Mössbauer evidence strongly supports the assignment: compounds 1–7 to an associated five-coordinate structure, and compounds 8–10 to an unassociated four-coordinate structure. Other spectroscopic evidence strongly supports this interpretation. This rather startling difference in structure is perhaps surprising considering the small difference in ligand between the compounds. For example, from compound 7 to compound 8, there is just replacement of a proton by a methyl group and yet the quadrupole splitting drops by over 1 mm s^{-1}.

Of greater general interest are the structures of the R$_3$SnX, Ph$_3$SnX, R$_2$SnX$_2$, and Ph$_2$SnX$_2$ compounds. The quadrupole splittings for some of these compounds are reported in Table 6.18. The X-ray structures of some

Table 6.18 Some Sn^{IV} quadrupole splitting data (mm s^{-1} at 80 K)

X	F	Cl	Br	I
Me$_3$SnX	3·82	3·44	3·39	3·10
Neo$_3$SnX†	2·79	2·65	2·65	2·40
Ph$_3$SnX	3·53	2·56	2·48	2·25
X	F	Cl	Br	I
Me$_2$SnX$_2$	4·38	3·55	3·36	—
Ph$_2$SnX$_2$	3·43	2·82	2·54	2·38

† Neo = (Me)$_2$CPhCH$_2$.
A. G. Maddock and R. H. Platt, J. Chem. Soc. (A), 1191 (1971); G. M. Bancroft and R. H. Platt, reference 6.1.

of these compounds have been reported. Me$_3$SnF, Me$_3$SnCl, Me$_2$SnF$_2$, and Me$_2$SnCl$_2$ all have associated structures at room temperature with five-coordinate (Me$_3$SnF, Me$_3$SnCl) and six-coordinate (Me$_2$SnF$_2$, Me$_2$SnCl$_2$) Sn atoms. Ph$_2$SnCl$_2$ and Ph$_3$SnCl contain four-coordinate Sn atoms as first predicted by Mössbauer and confirmed by X-ray. However, because of the difficulty in assigning a Van der Waals radius to Sn, there has been some dispute amongst the crystallographers about whether Me$_2$SnCl$_2$ is associated or not.

The quadrupole splitting of a known four-coordinate compound Ph$_2$ISn(CH$_2$)$_4$SnIPh$_2$ is 2·37 mm s^{-1}, and thus four-coordinate Ph$_3$SnX and R$_3$SnX should give quadrupole splittings of ~2·4 ± 0·3 mm s^{-1} with four-coordinate Me$_3$SnX compounds giving quadrupole splittings about 0·2 mm s^{-1} higher than Ph$_3$SnX quadrupole splittings [(p.q.s.)$_{Me}$ = −1·37 mm s^{-1}; (p.q.s.)$_{Ph}$ = −1·26 mm s^{-1}, Table 6.15]. Neo$_3$SnI and Ph$_3$SnI have very similar quadrupole splittings to Ph$_2$ISn(CH$_2$)$_4$SnI(Ph$_2$) (Table 6.18), and thus must have the unassociated tetrahedral configuration. Similarly, the compounds Neo_3SnX (X = F, Cl, Br) and Ph_3SnX (X = Cl, Br) give just slightly larger values as expected from the difference in partial quadrupole splittings for I and X(= F, Cl, Br)† in Table 6.15 and thus must also be four-coordinate. However, Ph_3SnF and Me_3SnX (X = F, Cl, Br, I) all have much larger quadrupole splittings, strongly indicating that they are all associated with five-coordinate Sn atoms.

For the Me_2SnX_2 and Ph_2SnX_2 compounds, it becomes more difficult to use the quadrupole splittings to distinguish between associated and unassociated structures. From Table 6.13, structures 2 and 3, the ratio of $(Q.S.)_{SnA_2B_2}:(Q.S.)_{SnAB_3} = (1 + \eta^2/3)^{1/2}:1 = 1.15:1$, and thus we would expect four-coordinate SnA_2B_2 compounds to have quadrupole splittings about 15% larger than their four-coordinate $SnAB_3$ counterparts. On this basis, four-coordinate structures can be assigned to Ph_2SnCl_2, Ph_2SnBr_2, and Ph_2SnI_2 (Table 6.18) with some confidence since the latter quadrupole splittings are ~10% higher than their Ph_3SnX counterparts. The four-coordinate structure for Ph_2SnCl_2 has been recently confirmed by X-ray crystallography (ref. 6.3). The very large quadrupole splitting for Me_2SnF_2 (4.38 mm s^{-1}) is consistent with its known associated structure. However, the quadrupole splittings for Me_2SnCl_2, Me_2SnBr_2, and Ph_2SnF_2 are in between those expected for associated and unassociated structures. Thus, the quadrupole splittings for Me_2SnCl_2 and Ph_2SnF_2 are 0.83 mm s^{-1} and 0.95 mm s^{-1} respectively lower than that for Me_2SnF_2, while a decrease of only about 0.3 mm s^{-1} would be expected if they both have associated six-coordinate structures. However the quadrupole splittings for Me_2SnCl_2, Ph_2SnF_2, and Me_2SnBr_2 are still well above those expected for four-coordinate structures, and it would appear from this evidence that these three compounds have weakly associated structures.

The quadrupole splittings thus have a great potential for studying intermolecular association (especially in conjunction with the temperature dependence of the recoil free fraction), but some caution should be exercised because of two effects. First, we assume ideal geometry in our partial quadrupole splitting calculations, and most structures are distorted. Distortions (as opposed to intermolecular association) could increase the expected quadrupole splitting. Second, we have been assigning $X_{intermolecular}$ the same partial quadrupole splitting as $X_{intramolecular}$, and yet $r_{Sn...X}$ is normally about 1 Å larger than r_{Sn-X}. However, the good agreement between predicted and observed quadrupole splittings strongly suggests that fairly large increases in quadrupole splitting (>0.5 mm s^{-1}) over that expected for an unassociated structure can be attributed to association.

The partial quadrupole splittings in Table 6.9 for Fe^{II} low spin compounds are also very useful for elucidating structure—for example whether a com-

† The larger quadrupole splittings for the fluorides suggest that the $(p.q.s.)_F$ is larger than that for $(p.q.s.)_{Cl,Br}$.

pound is *cis*- or *trans*-FeA$_2$B$_4$. Infrared and other spectroscopic techniques are often useful here, but sometimes Mössbauer appears to be the most useful spectroscopic means of assigning structure. The observed 2:1 ratio for the pairs of isomers in Table 6.5 enables an immediate assignment of the *trans*- and *cis*-isomers for these six compounds. The structure of FeClSnCl$_3$-(depe)$_2$ (compound #23, Table 6.8) can also be assigned using the partial quadrupole splittings in Table 6.9. The predicted quadrupole splitting for *trans*-FeClSnCl$_3$(depe)$_2$ is +1.02 mm s^{-1}; while for *cis*-FeClSnCl$_3$(depe)$_2$ the predicted value is just greater than −0.51 mm s^{-1} (due to a small η term). The observed value of 1.28 mm s^{-1} argues strongly for a *trans* structure in this case.

Similarly, the *trans*- and *cis*-isomers of Fe(NCS)$_2$(ArCN)$_4$ would have a predicted quadrupole splitting of +0.80 mm s^{-1} and −0.40 mm s^{-1}. The recent measurement of these quadrupole splittings confirms this prediction and allows the immediate assignment of these two structures. The structure of any compound containing the ligands in Table 6.9 (with the possible exception of strong π acceptor ligands such as CO or NO$^+$) could be assigned with confidence using the partial quadrupole splittings.

For other compounds of Fe^{II}, such as Fe(CO)$_2$X$_2$P$_2$ (X = Cl, Br, I; P = phosphines, phosphites) where there are five geometric isomers, the partial quadrupole splittings should be very useful for assigning structures. Two of the possible five isomers are given in Table 6.13 (species 12 and 13), and the EFG expressions indicate that the *trans* A isomer (species 12) should have a significantly larger quadrupole splitting than the all *cis* isomer (species 13). Present results indicate that these isomers can be assigned using partial quadrupole splittings, although the somewhat variable partial quadrupole splitting for CO can sometimes make the assignment somewhat ambiguous.

Bonding. In section 6.1, the variation of quadrupole splitting with electronegativity and bonding properties of ligands was discussed for a large number of compounds in which just one or two ligands were changed. The partial quadrupole splittings are very useful for unravelling trends in quadrupole splittings for series of compounds in which more extensive substitution takes place.

For Fe^{II} compounds, partial centre shifts (chapter 5, Table 5.12) decrease with increasing σ bonding and π back bonding ability of the ligand, while the partial quadrupole splittings become more positive with increasing π backbonding, but more negative with increasing q_lattice and increasing σ donor properties. Thus:

$$\text{p.c.s.} = -k(\sigma + \pi) \tag{5.7}$$

$$\text{p.q.s.} \propto -(pc)_L - (\sigma_L - \pi_L) \tag{6.18}$$

Mössbauer spectroscopy

These two equations provide a very useful method for characterizing both the σ donor and π acceptor properties of ligands. As expected from the above two equations, H^- and NO^+, the best σ donor and π acceptor ligands respectively give the most negative partial centre shifts of all ligands (Table 5.12) but have the most negative and positive partial quadrupole splittings respectively (Table 6.9). Considering some of the neutral ligands, the partial centre shifts show that $\sigma + \pi$ increases in the order $H_2O < NH_3 < depe/_2 \sim depb/_2 < dmpe/_2 < ArNC < CO$, while the partial quadrupole splittings show that $\pi - \sigma$ increases in the order $RNC < dmpe/_2 < depe/_2 < depb/_2 < CO < NH_3 < H_2O$. For the first five ligands in the above partial centre shift series, the σ donor properties probably dominate both the partial centre shifts and partial quadrupole splittings. A reasonably good correlation between partial centre shift and partial quadrupole splitting (Fig. 6.7) is

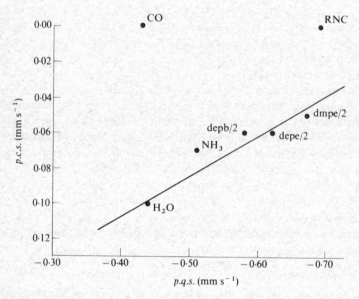

Fig. 6.7 Partial centre shift versus partial quadrupole splitting for ligands in Fe^{II} low spin compounds (Tables 5.12 and 6.9).

obtained with a positive slope, as for the Fe and Ru compounds noted earlier (Figs 6.3, 6.4, 6.5). However RNC, CO (and NO^+) lie well to the left of this line because the π acceptor properties are becoming more important in the order $RNC < CO < NO^+$. As discussed earlier in this chapter, if the π acceptor properties of a ligand dominate both partial quadrupole splittings and partial centre shifts, then a line of opposite slope to that in Fig. 6.7 would be observed. For charged ligands such as Cl^- and CN^-, it is more

144

Bonding and structure from quadrupole splittings

difficult to evaluate σ and π because of possible contributions to partial quadrupole splittings from $(pc)_L$.

For Sn^{IV} compounds, partial centre shifts generally cannot be derived, and so we cannot correlate partial centre shifts and partial quadrupole splittings. Consider partial quadrupole splittings (Table 6.15). Since it has been shown that π bonding will have no observable effect on the partial quadrupole splittings, the partial quadrupole splittings should become more negative as the σ donor capacity of L increases. Because the partial quadrupole splittings are only a measure of the relative p_z electron density, it is more appropriate here to say that the partial quadrupole splittings become more negative as the p_σ donor ability of L increases. Referring to the tetrahedral part of Table 6.15, p_σ decreases from R to NCS. This order is not always consistent with the σ donor order from centre shifts (chapter 5.1), and this apparent anomaly is due to the widely varying s characters of Sn–L bonds. For example, the metal moieties $Mn(CO)_5$ and $Fe(CO)_2Cp$ are poorer p donors than R and Ph (Table 6.15), yet the centre shifts show that the Sn–M bonds (M = $Mn(CO)_5$; $Fe(CO)_2Cp$) have a higher s character than the Sn–C bonds (section 5.1).

The signs of the quadrupole splittings for Sn^{IV} compounds can now be reexamined. For compounds such as *trans*-R_2SnX_4, the Z EFG axis lies along the four-fold molecular axis, and because R is a better p donor than X, p-electron density is concentrated along the Z EFG axis giving a negative V_{ZZ}, but a positive quadrupole splitting ($Q = -ve$). Similarly, for tetrahedral R_3SnCl compounds, the Z EFG axis lies through the three fold molecular axis. In this case, the electronegative ligands lie along the Z EFG axis and V_{ZZ} is positive (Q.S. = $-ve$). A negative quadrupole splitting is also expected for R_3SnX_2 compounds, where the electronegative X ligands are in the *trans* axial positions. All of these expectations have been confirmed by the magnetic field technique, although the positive signs for *cis*-R_2SnL_4 compounds noted earlier are not expected.

6.5 The effect of $q_{C.F.}$

In the Mössbauer species discussed in the last sections of this chapter, $q_{C.F.} = 0$, and quadrupole splittings were simply a function of the nature and distribution of the ligand bonds. For many other Mössbauer species, (e.g. Fe^{II} high spin, Fe^{III} low spin, Fe°, Fe^{-1}, Au^{III}, Ru^{III}) the $q_{C.F.}$ term (eq. 2.16) sometimes obscures the dependence of the quadrupole splitting on the nature of the M–L bonds. In other cases (Sn^{II}, Sb^{III}), the inherent asymmetric occupation of the valence orbitals (Sn^{II}, $Sb^{III} = 5s^{2-x}5p^x$) obscures the dependence of the quadrupole splitting on the M–L bond types. However, for Fe^{II} high spin and Fe^{III} low spin compounds in which $q_{C.F.}$ dominates the $q_{valence}$ contribution, there has been considerable success in interpreting quadrupole splittings and their temperature dependence. For many other

Mössbauer spectroscopy

Mössbauer species such as Fe° and Fe^{-1}, interpretation of the quadrupole splittings—except on the purely empirical level—has usually not been possible. In these cases, $q_{M.O.}$ and $q_{C.F.}$ can no longer be realistically or usefully separated.

In this section, the Fe^{II} quadrupole splitting treatment (ref. 6.5) will be outlined, followed by structural and bonding uses for Fe^{II} high spin compounds. Then, Fe° and Au^{III} quadrupole splittings will be briefly discussed.

Fe^{II} high spin. As discussed in Chapter 2, if Fe^{2+} high spin ($t_{2g}^4 e_g^2$) is surrounded by a perfect octahedral crystal field, then the degeneracy of the e_g and t_{2g} orbitals is not removed and $q_{C.F.} = 0$. However, Fe^{2+} high spin is inherently subject to a Jahn–Teller distortion (Fig. 6.8), which removes the degeneracy

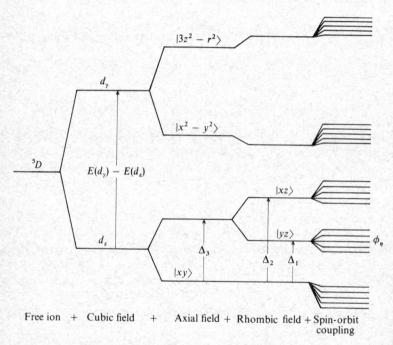

Fig. 6.8 Energy level diagram for Fe^{2+} high spin [R. Ingalls, *Phys. Rev.*, **133A**, 787 (1964).]

of the t_{2g} and e_g orbitals, and gives rise to a large $q_{C.F.}$ term. If we consider that there is a large enough axial compression in the octahedral case such that the fourth t_{2g} electron exclusively occupies the low energy d_{xy} orbital, then $q_{C.F.} = 4/7 \langle r^{-3} \rangle_{3d}$, which should give a quadrupole splitting of about 4 mm s^{-1} (chapter 2.2). However, $q_{valence}$ is reduced from this value by: a)

146

Boltzmann population of the other t_{2g} levels; b) spin-orbit coupling; c) covalency effects and d) $q_{M.O.}$, which is partially taken into account by the covalency. In addition, the $q_{lattice}$ term opposes $q_{C.F.}$ as noted earlier in chapter 2. $q_{valence}$ can now be written (ref. 6.5):

$$q_{valence} = 4/7 \langle r^{-3} \rangle_0 \alpha^2 F(\Delta_1, \Delta_2, \alpha^2 \lambda_0, T) \tag{6.23}$$

where: the function F expresses the decrease in $q_{valence}$ due to thermal population and spin-orbit coupling,
λ_0 is the spin-orbit coupling constant in the free ion
and, $\alpha^2 = \langle r^{-3} \rangle / \langle r^{-3} \rangle_0$ is the covalency parameter and usually takes the values 0·6 to 0·9. This parameter expresses the extent of radial expansion of the $3d$ orbitals.

Neglecting spin-orbit coupling, and taking $\Delta_1 = \Delta_2 = \Delta_3$ (Fig. 6.8) then:

$$F(\Delta_3, T) = \frac{1 - \exp(-\Delta_3/kT)}{1 + 2\exp(-\Delta_3/kT)} \tag{6.24}$$

$F \to 1$ as $T \to 0$ K. If, instead, the orbital doublet is lowest in energy:

$$F(\Delta_3, T) = \frac{1 - \exp(-\Delta_3/kT)}{2 + \exp(-\Delta_3/kT)} \tag{6.25}$$

$F \to 1/2$ as $T \to 0$ K. For tetrahedral compounds, in which the d_{z^2} and $d_{x^2-y^2}$ orbitals are lowest in energy, then:

$$F(\Delta_3, T) = \frac{1 - \exp(-\Delta_3/kT)}{1 + \exp(-\Delta_3/kT)} \tag{6.26}$$

$F \to 1$ as $T \to 0$ K.

The spin-orbit coupling lifts the five fold degeneracy of each orbital state (Fig. 6.8). This decreases F by an amount depending on the ratios Δ_1/λ and Δ_2/λ. The quadrupole splitting is also decreased by the expansion of the radial part of the $3d$ wave function on bonding. The covalency parameter takes this into account.

The lattice contribution for the tetragonal distortions to the octahedral case can be expressed as (ref. 6.5):

$$q_{lattice} = \pm \frac{14\Delta_3}{3e^2 \langle r^2 \rangle} \tag{6.27}$$

As mentioned in chapter 2, $q_{lattice}$ has the opposite sign to $q_{valence}$. At a given temperature, the quadrupole splitting increases to about 4 mm s^{-1} as Δ_3 increases (mostly due to $q_{valence}$); the quadrupole splitting then decreases with increasing Δ_3 as $q_{lattice}$ increases (Fig. 6.9).

The $q_{lattice}$ term, spin-orbit coupling, covalency and $q_{M.O.}$ have often been neglected, and the quadrupole splitting as a function of temperature has

Mössbauer spectroscopy

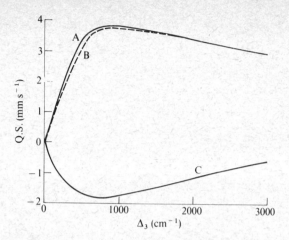

Fig. 6.9 Plot of quadrupole splitting against Δ_3; A, d_{xy} lowest energy; B, as in A but allowing for spin orbit coupling; C, d_{xz}, d_{yz} lowest energy. [C. D. Burbridge, D. M. L. Goodgame, and M. Goodgame, *J. Chem. Soc.* (A), 349 (1967).]

been fitted using eqs (6.24), (6.25), or (6.26) in the form:

$$\text{Q.S.} = (\text{Q.S.})_0 F(\Delta_3, T) \qquad (6.28)$$

where $(\text{Q.S.})_0$ is the quadrupole splitting at $T = 0$ K. With quadrupole splittings at two temperatures, both Δ_3 and $(\text{Q.S.})_0$ can be solved for. Fits to the Q.S.(T) curves over an extended temperature range have generally not been entirely satisfactory due to the neglect (usually) of q_{lattice}, $q_{\text{M.O.}}$ and spin-orbit coupling. $(\text{Q.S.})_0$ in eq. (6.28) takes the direct dependence of quadrupole splitting on α^2 (eq. 6.23) into account. For Fe^{III} low spin compounds, the agreement between the calculated and observed Q.S.(T) curves has been considerably less satisfactory than for Fe^{II} high spin compounds.

However, the measurement of Fe^{II} quadrupole splittings as a function of temperature is important in at least three ways: first, to detect small, otherwise undetectable, distortions from perfect tetrahedral or octahedral symmetry; second, to estimate the crystal field splittings Δ_1, Δ_2 or Δ_3; and third, to estimate the covalency factor α^2 (ref. 6.6). To consider the first two points, the large and temperature-dependent quadrupole splitting for $FeCl_4^{2-}$ and $Fe(NCS)_4^{2-}$ compounds (Table 6.19) indicate that these formally tetrahedral species are slightly distorted from tetrahedral symmetry. These small distortions had not been detected by X-ray crystallography. Using eqs (6.26) and (6.28), the magnitude of Δ_3 was obtained as in Table 6.20. The orbital ground state in these compounds was obtained by taking magnetic spectra of $(NMe_4)_2FeCl_4$. The low-energy line split into two, and the high-energy line into three. V_{zz} and q are therefore negative and the lowest energy orbital

Table 6.19 Quadrupole splittings for Fe^{II} high spin compounds (mm s^{-1})

Compound	T(K)	Q.S.	Compound	T(K)	Q.S.
(NMe$_4$)$_2$FeCl$_4$	4	3.27	(NMe$_4$)$_2$Fe(NCS)$_4$	4	2.83
	77	2.61		77	2.10
	293	0.72	Fe(IQ)$_4$Cl$_2$†	295	3.18
(NEt$_4$)$_2$FeCl$_4$	4	3.21	Fe(IQ)$_4$Br$_2$	295	2.21
	77	2.68	Fe(IQ)$_4$I$_2$	295	0.40
	293	1.16			

† IQ = isoquinoline.
P. R. Edwards, C. E. Johnson and R. J. P. Williams, *J. Chem. Phys.* **47**, 2074 (1967); C. D. Burbridge, D. M. L. Goodgame and M. Goodgame, *J. Chem. Soc.* (A), 349 (1967).

is d_{z^2}. This ground state corresponds to a distortion having the form of a compression of the tetrahedron along the Z axis (ref. 6.7).

Many other Fe^{II} quadrupole splittings have been obtained just at room temperature, and Δ_3 values calculated assuming a value of (Q.S.)$_0$ in eq. (6.28). For Fe(IQ)$_4$X$_2$ (IQ = isoquinoline, X = Cl$^-$, Br$^-$, I$^-$) (Table 6.20) the quadrupole splittings were used with eqs (6.24) and (6.28) to calculate the Δ_3 values in Table 6.20. The signs of these splittings are positive, and d_{xy} must then be the ground state. It has recently been shown that Q.S.(T) for such octahedral compounds does not fit eqs (6.24) and (6.28) very well—presumably due to the neglect of spin-orbit coupling and q_{lattice}.

In an attempt to sort out the importance of the various factors involved in q_{valence} and the quadrupole splitting in Fe^{II} compounds, it has recently been suggested (ref. 6.6) that the major source of differences in quadrupole splittings for similar series of compounds with similar $F(\Delta, T)$ is associated

Table 6.20 Crystal field splittings in Fe^{II} high spin compounds

Compound	Δ_3(cm^{-1})	Ground state orbital
(NEt$_4$)$_2$FeCl$_4$	135	d_{z^2}
(NMe$_4$)$_2$FeCl$_4$	125	d_{z^2}
(NMe$_4$)$_2$Fe(NCS)$_4$	101	d_{z^2}
Fe(IQ)$_4$Br$_2$†	360	d_{xy}
Fe(IQ)$_4$I$_2$	80	d_{xy}
Fe(IQ)$_4$Cl$_2$	600	d_{xy}

† IQ = isoquinoline.
P. R. Edwards, C. E. Johnson and R. J. P. Williams, *J. Chem. Phys.*, **47**, 2074 (1967); C. D. Burbridge, D. M. L. Goodgame and M. Goodgame, *J. Chem. Soc.* (A), 349 (1967).

Mössbauer spectroscopy

with differences in α^2. Using the value of $\alpha^2 = 0.60$ for FeF_2 from e.s.r. data, α^2 was determined to be 0.34, 0.39, and 0.42 for the iodide, bromide, and chloride respectively. This order parallels the nephelauxetic series. Also, as noted in chapter 5, the variation of centre shift for these compounds shows an excellent correlation with the nephelauxetic series in contrast to the correlation with the spectrochemical series for Fe^{II} low spin and Au^{III} centre shift values.

Similarly for $[FeCl_{6-n} \cdot nH_2O]^{n-4}$ compounds, plots of quadrupole splitting and isomer shift (Fig. 5.6) versus n are reasonably linear, and the correlation of quadrupole splitting with isomer shift has been attributed to central field covalency—the expansion of the radial portion of the $3d$ wave function (causing a decrease in α^2) due to the reduction of the metal ion's effective charge via σ and π bonding. σ and π bonding properties of ligands should be forthcoming from such correlations.

Fe° and π-Cp Fe compounds. An enormous number of Fe° and π-Cp Fe (Cp = C_5H_5) quadrupole splittings have been reported. A very small number of quadrupole splittings are given in Table 6.21. Variations in quadrupole splitting for fairly similar groups of compounds have not been amenable to any consistent qualitative or semiquantitative treatment—often due to the large and variable $q_{C.F.}$ term. For example, for $Fe(CO)_2CpX$ (X = Cl, $SnCl_3$,

Table 6.21 Quadrupole splittings for Fe° compounds and π-Cp Fe compounds (mm s^{-1}, usually at 80 K)

Compound	Q.S.
1. $Fe(CO)_2CpCl$	1.88
2. $Fe(CO)_2CpSnCl_3$	1.82
3. $Fe(CO)Cp(PPh_3)SnPh_3$	1.84
4. $(Cp)_2Fe$†	2.39
5. $[Cp_2Fe]^+Br^-$	~0.2
6. $Fe(CO)_5$	2.57
7. $(Et_4N)_2[Fe_2(CO)_8]$	2.22
8. $(Et_4N)[Fe(CO)_4H]$	1.36
9. $Na_2Fe(CO)_4$	<0.18
10. $Fe_2(CO)_9$	0.48
11. $(Et_4N)[Fe_2(CO)_8H]$	0.50
12. $[Fe(CO)_3(PMe_2)]_2$	0.69

† Cp = π − C_5H_5.
R. Greatrex and N. N. Greenwood, *Disc. Faraday Soc.*, **47**, 126 (1969). References in: G. M. Bancroft and R. H. Platt, *Adv. Inorg. Radiochem.*, **15**, 59 (1972); ed. H. J. Eméleus and A. G. Sharpe, Academic Press.

$SnPh_3$, GeX_3, SiX_3, etc.) the quadrupole splittings are remarkably constant at about 1.8 mm s^{-1}, (Table 6.21) despite the widely varying bonding properties of the ligands. There is probably a large variation in π bonding between the Fe and CO groups which fortuitously masks any changes in quadrupole splitting due to the different bonding properties of the varied ligand.

The very large quadrupole splitting in $(Cp)_2Fe$ and the very small quadrupole splitting in $[Cp_2Fe]^+Br$ (Table 6.21) has been the subject of considerable discussion. The sign of the quadrupole splitting in $(Cp)_2Fe$ has been found to be positive, and this sign has been attributed mainly to an electron localized in d_{xy}. The removal of this electron in $[(Cp)_2Fe]^+$ complexes fortuitously collapses the quadrupole splitting.

In addition, there have been several useful empirical generalizations of Fe carbonyl quadrupole splittings (ref. 6.8 and Table 6.21). First, five-coordinate Fe° compounds such as $Fe(CO)_5$, $Fe_2(CO)_8$ and $Fe(CO)_4H$ have much larger quadrupole splittings than four-coordinate compounds such as $Na_2Fe(CO)_4$ and six-coordinate Fe° compounds such as $[Fe(CO)_3(PMe_2)]_2$ (including an Fe–Fe bond). Seven-coordinate compounds such as $Fe_2(CO)_9$ and $Fe_2(CO)_8H^-$ also have small quadrupole splittings generally. However, beyond these purely empirical generalizations, there has been little progress in rationalizing other Fe°, Fe^{-I}, and intermediate spin Fe^{III} quadrupole splittings.

Au^{III} compounds. Au^{III} (d^8) compounds are known to be square planar, and σ bonding involves dsp^2 hybrid orbitals on the gold. π bonding has been thought to be minimal in the compounds discussed below, and it is neglected. The Z EFG axis lies along the four-fold symmetry axis. The $d_{x^2-y^2}$ hole in the Au^{3+} ion would then produce a negative $q_{C.F.}$ and e^2qQ (Q is positive), since there is a concentration of electron density along the Z EFG axis. Covalent bonding, however, to the $5d_{x^2-y^2}$, $6p_x$ and $6p_y$ orbitals produces a positive $q_{M.O.}$ (eqs 2.14 and 2.15), and except for the most ionic Au^{III} compounds (such as AuF_4^-), a positive e^2qQ should result ($|q_{M.O.}| > |q_{C.F.}|$). Supporting these assignments, a plot of centre shift versus quadrupole splitting for a number of Au^{III} compounds (ref. 6.9) is reasonably linear with an equation:

$$\text{C.S.} = 0.542 \, \text{Q.S.} + 0.016 (\text{cm s}^{-1}). \tag{6.29}$$

Like the centre shift then, the quadrupole splitting is mainly dependent on the σ donor properties of the ligands and both become more positive as the σ donor properties of the ligand increase. Thus the relatively ionic $AuCl_4^-$ gives the most negative centre shift and quadrupole splitting, (Table 6.22) while the covalent $Au(CN)_4^-$ gives the most positive centre shift and quadrupole splitting. Unlike Fe° quadrupole splittings, Au^{III} quadrupole splittings show

Mössbauer spectroscopy

Table 6.22 Quadrupole splittings for Au(III) compounds (mm s^{-1} at 4 K)

Compound	Q.S.
KAuI$_4$	(+)1·28
KAuBr$_4$	(+)1·13
KAuCl$_4$	(+)1·11
KAu(SCN)$_4$	(+)2·04
As(C$_6$H$_5$)$_4$Au(N$_3$)$_4$	(+)2·89
Na$_3$AuO$_3$	(+)3·02
KAu(CN)$_2$Br$_2$	(+)5·34
KAu(CN)$_4$	(+)6·86

H. D. Bartunik, W. Potzel, R. L. Mössbauer and G. Kaindl, *Z. Physik*, **240**, 1 (1970).

definite trends which can be rationalized in terms of σ bonding properties of the ligands. The $q_{C.F.}$ contribution seems to be reasonably constant for the Au^{III} compounds with variations in quadrupole splitting being mainly due to $q_{M.O.}$.

Problems

1. How would you expect the quadrupole splitting to vary in the following compounds: (consider that all are tetrahedral) Ph$_4$Sn, Ph$_3$SnCl, Ph$_2$SnCl$_2$, PhSnCl$_3$ and Cl$_4$Sn?
2. How would you expect the quadrupole splittings to vary about the asterisked atoms in the following pairs of compounds?
   (a) tetrahedral and square planar *SnCl$_4$
   (b) *trans-* and *cis-*Fe(CO)$_4$Cl$_2$
   (c) high spin *trans-* and *cis-*Fe(NH$_3$)$_4$Cl$_2$
   (d) tetrahedral Me$_3$*SnCl and five-coordinate (Cl axial) Me$_3$*SnCl
   (e) isostructural Me$_3$*SnCl$_2^-$ and Me$_3$*SbCl$_2$.
3. Consider that the SnCl$_3^-$ anion in MSnCl$_3$ compounds (M = Na$^+$, K$^+$ etc.) has a Cl–Sn–Cl bond angle of 90° (implying that the lone pair of electrons is an "inert" 5s pair). Using the point charge model, what is the expected quadrupole splitting for this idealized SnCl$_3^-$ ion? How would you expect the quadrupole splitting and centre shift to vary as the Cl–Sn–Cl bond angle increases to the tetrahedral angle 109°28'?
4. How many geometric isomers are there in compounds of the type MA$_2$B$_2$C$_2$ and MAB$_3$C$_2$? Calculate the point charge EFG expressions for each structure. Can the Z EFG axis be defined for each of these

structures without knowing the sign and magnitudes of the partial quadrupole splittings?

5. Calculate the quadrupole splittings expected for the above structures given the following partial quadrupole splittings.
   (p.q.s.)$_A$ = -0.70 mm s^{-1} (could be ArNC)
   (p.q.s.)$_B$ = -0.55 mm s^{-1} (could be CO)
   (p.q.s.)$_C$ = -0.30 mm s^{-1} (could be Cl)
   Label the EFG axes. What happens to the quadrupole splitting when (p.q.s.)$_A$ = (p.q.s.)$_B$?

6. The compound BuSnCl$_3$(py)$_2$ (Bu = n–butyl, py = C$_5$H$_5$N) gives an observed quadrupole splitting of 1.86 mm s^{-1}. Using the point charge expressions in problem 2 and the partial quadrupole splittings for R, Cl and py in Table 6.15, calculate the expected quadrupole splittings for the three possible isomers. Can you assign the observed quadrupole splitting to one of the three structures with confidence?

7. A plot of $(e^2qQ)_{121Sb}$ versus $(e^2qQ)_{119Sn}$ (both in mm s^{-1}) for isoelectronic Sn and Sb compounds of the type R$_3$MX$_2$ (R=Me, Ph; M=Sn, Sb; X=Cl, Br, I) gives a slope of $+3.40$ with zero intercept. Taking Q_{121Sb} = -0.28 b, $q_{5p(Sn)} = 11.2a_o^{-3}$, $q_{5p(Sb)} = 13.0a_o^{-3}$, $E_\gamma(Sb) = 37.2$ keV and and $E_\gamma(Sn) = 23.9$ keV, calculate the quadrupole moment of ^{119}Sn.

8. The species [Fe^{II}H(dppe)$_2$]$^+$ has been recently prepared (P. Giannoccaro, M. Rossi and A. Sacco, *Coord. Chem. Rev.*, **8**, 77, (1972)). Assuming that the structure is a square pyramid, calculate the expected quadrupole splitting for this Fe^{II} low spin species taking (p.q.s.)$_{dppe/2}$ = -0.59 mm s^{-1} and the value for H$^-$ given in Table 6.9. Is this quadrupole splitting dependent on the partial quadrupole splitting of chloride?

9. The compounds [FeHL(dppe)$_2$]$^+$A$^-$ (A$^-$ = ClO$_4$, etc.) have recently been made. How would you expect the Mössbauer parameters to compare with those for the analogous [FeHL(depe)$_2$]$^+$ compounds? Take (p.q.s.)$_{dppe/2}$ = -0.59, and (p.c.s.)$_{dppe/2}$ = $+0.08$ and the values for depe in Tables 5.12 and 6.9. Using chemical intuition and the partial quadrupole splitting values in Table 6.9, how would you expect the quadrupole splittings and centre shifts to vary for L = (CH$_3$)$_2$CO, NH$_3$, N$_2$ and CO?

10. Using the partial quadrupole splitting values given in Table 6.15, calculate the expected quadrupole splitting for tetrahedral PhSnCl$_3$. Compare this with the observed value, and preliminary calculated value in Table 6.14. Calculate the expected quadrupole splitting for an octahedral structure (two chlorines bridging). What is the likely structure of this compound?

11. In compounds containing an L–M–I linkage (L = neutral ligand, M = transition metal), how would you expect the ^{129}I quadrupole splitting to be effected as L becomes a stronger σ donor?

References

6.1 G. M. Bancroft and R. H. Platt in *Advances Inorganic and Radiochemistry*, ed. H. J. Emeleus and A. G. Sharpe, Academic Press **15**, 59 (1972).
6.2 M. G. Clark, A. G. Maddock and R. H. Platt, *J. Chem. Soc.* (Dalton), 281 (1972).
6.3 P. T. Greene and R. F. Bryan, *J. Chem. Soc.* (*A*), 2549 (1971).
6.4 R. H. Herber, *J. Chem. Phys.*, **54**, 3755 (1971), and references.
6.5 R. Ingalls, *Phys. Rev.*, **133A**, 787 (1964).
6.6 Y. Hazony and R. C. Axtmann, *Chem. Phys. Letts.*, **8**, 571 (1971) and references.
6.7 P. R. Edwards, C. E. Johnson and R. J. P. Williams, *J. Chem. Phys.*, **47**, 2074 (1967).
6.8 K. Farmery, M. Kilner, R. Greatrex, and N. N. Greenwood, *J. Chem. Soc.* (*A*), 2339 (1969).
6.9 M. O. Faltens and D. A. Shirley, *J. Chem. Phys.*, **53**, 4249 (1970).

7. Mössbauer spectroscopy as a fingerprint technique in mineralogy and geochemistry

The mineralogical and geochemical applications of Mössbauer spectroscopy have developed rapidly over the last few years. Although a few spectra of tin and tungsten minerals have been recorded, the vast majority of spectra reported have been iron spectra, mainly because of the wide and varied occurrence of iron in the earth's crust. Most of the applications in this area can be considered fingerprint applications of varying degrees of sophistication. Much of the information obtainable from Mössbauer spectra cannot be obtained using other techniques.

In this chapter, we will discuss some of the more qualitative fingerprint applications: characterization of the oxidation state of iron (e.g., Fe^{2+} or Fe^{3+}), electronic configuration of iron (e.g., high or low spin), coordination symmetry about the iron atom (e.g., tetrahedral or octahedral) and site distortion from either octahedral or tetrahedral symmetry; assignment of the peaks to structurally distinct cation positions; correlation of the quadrupole splitting values with structural variations; and perhaps most importantly, qualitative uses of the previous assignments for explaining solid-state processes such as oxidation and weathering of minerals. The other major, more quantitative, applications centre around the quantitative estimation of iron in structurally distinct iron positions in one phase, or semiquantitative estimation of iron containing minerals in bulk samples such as lunar and meteoritic materials. These latter two applications will be discussed in the ensuing two chapters.

As noted in section 3.3, mineral spectra often consist of a number of strongly overlapping lines; and detailed computing is essential to obtain a satisfactory interpretation. The well-behaved Lorentzian line shapes enable detailed computing. In addition, most mineral spectra give line widths of ~ 0.30 mm s^{-1} with a narrow line source, and it is reasonable then to constrain line widths of component peaks to be equal. Often, the areas of component doublet peaks are equal for random samples, and the areas may also be constrained to be equal. Deviations from the above equalities will be noted in the following sections, and then line widths and areas will be further discussed in section 7.4.

7.1 Oxidation state and electronic configuration of iron in minerals

In Table 4.1, we saw that the Mössbauer centre shift is sensitive to the high spin oxidation states of iron. For example, Fe^{+6} and Fe^{+1} typically have centre shift values of -0.6 and $+2.2$ mm s^{-1}, respectively, while the mineralogically important Fe^{2+} and Fe^{3+} have characteristic centre shift values of $+1.4$ mm s^{-1} and $+0.7$ mm s^{-1} respectively in an oxygen environment. These values vary somewhat depending on the coordination number, site symmetry, and type of ligand (e.g., O or S) (Table 7.1 and Fig. 7.1), but normally there is little or no ambiguity in assigning the oxidation state of iron in minerals from the Mössbauer centre shift.

The assignment of the oxidation state of iron, or the ratio of oxidation states, is of considerable geological importance in explaining such phenomena as colour, pleochroism, oxidation, and weathering of minerals. These will be discussed in section 7.6. In addition, the assignment of oxidation states may be of great importance as a geothermometer and geobarometer. For example, Drickamer's work, section 4.4, shows that Fe^{3+} is reduced to Fe^{2+} with pressure, and thus the ratio of these ions in the earth's interior may form the basis of a useful geobarometer. For some mineral series, such as pyroxenes, the Fe^{2+}/Fe^{3+} ratio has been correlated with the temperature and/or pressure of formation and the oxygen pressure during crystallization.

The determination of oxidation state is not a trivial problem, because standard techniques such as X-ray crystallography or the electron probe cannot distinguish between Fe^{2+} and Fe^{3+}, and chemical analyses sometimes give unreliable values for these two ions, especially when the mineral is difficult to dissolve, or when there are other transition metal ions present which are easily oxidized or reduced.

A number of examples illustrate the success of the Mössbauer method. Two simple examples involve iron in a mineral with another element, such as antimony, tellurium, or titanium, which exhibits variable valency. The usually accepted formula for tripuhyite is $Fe_2Sb_2^VO_7$ (or $2FeO \cdot Sb_2^VO_5$), thus implying that iron is in the ferrous state. However, later X-ray studies and chemical analyses suggested that the ideal formula is $FeSb^VO_4$, implying that iron is in the ferric state. The Mössbauer spectrum of tripuhyite (ref. 7.1) gave a single quadrupole doublet with a centre shift of 0·62 mm s^{-1} and a quadrupole splitting of 0·72 mm s^{-1}. The centre shift value is characteristic of ferric iron (Table 7.1) and proves conclusively that at least 95% of the iron is in the ferric state and that the simplest chemical formula should be

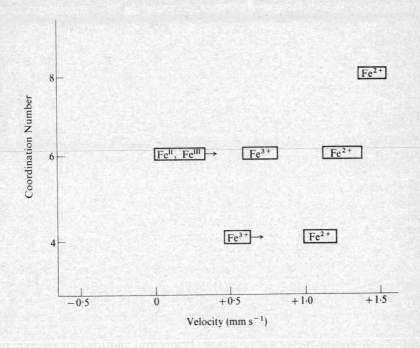
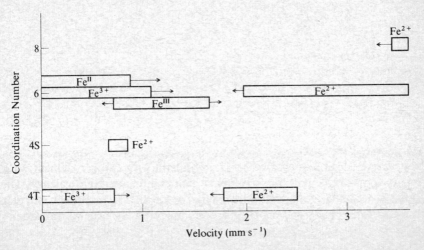

Fig. 7.1 ^{57}Fe Centre shifts (top) and Quadrupole Splitting (bottom) plotted versus coordination number for 'ionic' high spin and low spin compounds and minerals. Arrows indicate that values outside the boxed areas have been observed. Adapted from G. M. Bancroft, A. G. Maddock and R. G. Burns, *Geochim. Cosmochim. Acta.*, 31, 2219 (1967).

Mössbauer spectroscopy

Table 7.1 Mössbauer centre shifts (mm s^{-1}) for various electronic configurations and coordination numbers of iron (at 295 K)

Mineral or species	Type of iron	C.S.
Almandine garnet	8 coordinate Fe^{2+}	1.56
Silicates	6 coordinate Fe^{2+}	1.30–1.43
Staurolite	4 coordinate Fe^{2+}	1.22
Spinels	4 coordinate Fe^{2+}	1.07
Gillespite	4 coordinate (square planar) Fe^{2+}	1.01
Chemicals (chapter 5)	low spin Fe^{II}	<0.7
Epidote	6 coordinate Fe^{3+}	0.61
Amphiboles	6 coordinate Fe^{3+}	~0.65
$FeCl_6^{\equiv}$	6 coordinate Fe^{3+}	0.76
$FeCl_4^-$	4 coordinate Fe^{3+}	0.56
Iron orthoclase	4 coordinate Fe^{3+}	0.72
$R_3Fe_2(FeO_4)_3$	⎧ 6 coordinate Fe^{3+}	~0.65
(R = rare earth)	⎩ 4 coordinate Fe^{3+}	~0.45
Chemicals (chapter 5)	low spin Fe^{III}	<0.6

G. M. Bancroft, A. G. Maddock, and R. G. Burns, *Geochim. Cosmochim. Acta*, **31**, 2219 (1967); M. J. Rossiter, *Phys. Lett.*, **21**, 128 (1966); G. M. Bancroft, M. J. Mays, and B. E. Prater, *J. Chem. Soc.*, (A), 956 (1970); F. F. Brown and A. M. Pritchard, *Earth Plan. Sci. Lett.*, **5**, 259 (1969); A. Hudson and H. J. Whitfield, *Mol. Phys.*, **12**, 165 (1967); W. J. Nicholson and G. Burns, *Phys. Rev.* **133**, A1568 (1964); G. M. Bancroft and R. G. Burns, *Min. Soc. Amer. Spec. Pap.* **2**, 137 (1969).

$Fe^{3+}SbO_4$. The figure of 95% is quoted because there may be a very small amount of Fe^{2+} which might not be seen in the Mössbauer spectrum due to the statistical scatter of the baseline.

Another example concerns the rare mineral neptunite, $LiNa_2K$ (Fe, Mn, Mg)$_2$ Ti_2O_2 (Si_8O_{22}), which has Ti in one octahedral position, and (Fe, Mn) in another slightly larger octahedral position. Although Ti should enter the octahedron with the smaller average metal–oxygen distance, because of the small difference in average radii of the two sites, it is possible that neptunite might contain Ti^{3+} and (Fe^{3+}, Mn^{3+}) ions of similar radii, rather than Ti^{4+} and (Fe^{2+}, Mn^{2+}) of very different radii. The Mössbauer spectrum of neptunite (ref. 7.2) resolves this question immediately. The spectrum consists of a single doublet having a centre shift of 1.42 mm s^{-1} and a quadrupole splitting of 2.65 mm s^{-1}. This spectrum conclusively proves that at least 95% of the iron is in the Fe^{2+} state, and indicates that titanium is present as Ti^{4+}.

Many minerals give spectra which indicate that appreciable amounts of both Fe^{2+} and Fe^{3+} are present. For example, the approximate chemical

Mössbauer spectroscopy as a fingerprint technique

formula of howieite is $NaFe_7^{2+}Mn_3Fe_2^{3+}Si_{12}O_{31}(OH)_{13}$, and the Mössbauer spectrum (Fig. 7.2) consists of two doublets: A and A' having a centre shift of 1.43 mm s^{-1}, and B and B' having a centre shift of 0.65 mm s^{-1}. The component peaks of a doublet are of identical width, but slightly different area, i.e. A and B are more intense than A' and B' respectively. From these centre shift values and Figure 7.1a, it is apparent that A and A' are due to Fe^{2+} and B and B' are due to Fe^{3+}. The ratio of areas A_A/A_B is $\sim 3:1$ in qualitative agreement with the above chemical formula ratio of 7:2.

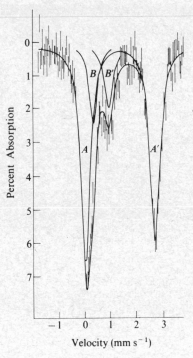

Fig. 7.2 Room temperature Mössbauer spectrum of howieite. [G. M. Bancroft, R. G. Burns, and A. J. Stone, *Geochim. Cosmochim. Acta*, **32**, 547 (1968).]

Figure 7.2 is typical of many other more complex spectra of minerals containing both Fe^{2+} and Fe^{3+}. The high velocity Fe^{3+} peak (B' in Fig. 7.2) is almost always fairly well resolved from the low velocity Fe^{2+} peak(s); however, peak B usually overlaps the low velocity Fe^{2+} peak(s) and is not usually resolved. There are three major reasons why we can be confident that peak B is present. First, the width and height of peak A would be much larger than peak A' without B present, and as discussed later, component peaks of a doublet normally have very similar widths and areas. Second, peak B' on its own gives an unrealistic centre shift of 0·95 mm s^{-1}—much higher than any known Fe^{3+} centre shift. Third, the addition of peak B to the spectrum always results in a large decrease in χ^2. However, the position, width, and area of peak B usually have a large standard deviation, making the centre shift and quadrupole splitting of doublet B and B' considerably less accurate than that of A and A'. In more complex spectra containing a number of Fe^{2+} doublets, it is often necessary to constrain the width and area of B to be equal to B' in order to obtain a fit.

Similarly, in sulphide minerals, the oxidation state of iron can usually be identified, although there are sometimes difficulties which we will consider later. Troilite (FeS) and pyrrhotite (Fe$_7$S$_8$) both have centre shifts of about 1·1 mm s^{-1} (Table 7.2), characteristic of Fe^{II} high spin. It should be noted that

Table 7.2 Mössbauer centre shifts (mm s^{-1}) for sulphide minerals

Mineral	C.S.	Temperature (K)	Assignment
Fe$_7$S$_8$ (pyrrhotite)	1·03–1·08	77	Fe^{II} high spin
FeS (troilite)	1·02	300	Fe^{II} high spin
Fe$_{1\cdot04}$S (mackinawite)	0·5	4·2	Fe^{II} low spin
Fe$_3$S$_4$ (greigite)	0·96	4·2	Fe^{II} high spin
	0·66		Fe^{III} high spin
	0·71		Fe^{III} high spin
(Ni, Fe)$_9$S$_8$ (pentlandite)	0·83	300	Fe^{II} high spin?
	0·62		Fe^{II} high spin?

D. J. Vaughan and M. S. Ridout, *Solid State Comm.*, **8**, 2165 (1970); S. Hafner and M. Kalvius, *Z. Krist.*, **123**, 443 (1966); D. J. Vaughan and M. S. Ridout, *J. Inorg. Nucl. Chem.*, **33**, 741 (1971).

this value is much smaller than the centre shifts for Fe^{2+} in oxygen coordination, but this can be attributed to the greater covalency of the Fe–S bond compared to that of the Fe–O bond.

Fe^{II} low spin compounds normally give much smaller centre shifts than Fe^{II} high spin compounds (Table 7.1, and chapter 5) and are normally comparable to, or slightly smaller than, Fe^{III} high-spin values. Surprisingly, the non-stoichiometric sulphide mackinawite, Fe$_{1\cdot04}$S, has a centre shift of 0·5 mm s^{-1} at 4·2 K, indicating that the iron is present as Fe^{III} high spin or Fe^{II} low spin. Application of a magnetic field to the sample indicated that it was diamagnetic, thus strongly suggesting that Fe^{II} low spin is present (section 4.4), in contrast to Fe^{II} high spin in troilite (FeS). To explain this rather surprising difference, it has been proposed that the d electrons in mackinawite may be extensively delocalized in the basal plane, forming metallic bands.

Several workers have noted that Fe^{II} high spin should revert to Fe^{II} low spin at very high pressures, and there is spectral evidence suggesting that in gillespite ($BaFeSi_4O_{10}$) (ref. 7.3), a high spin–low spin transition takes place at high pressures. From magnetic susceptibility evidence, one mineral, deerite (approximate formula $Fe_{13}^{2+}Fe_7^{3+}Si_{13}O_{44}(OH)_{11}$) appeared to contain a large fraction (>50%) of its iron in a low spin state. The Mössbauer spectrum of deerite at room temperature is shown in Fig. 7.3. It consists of

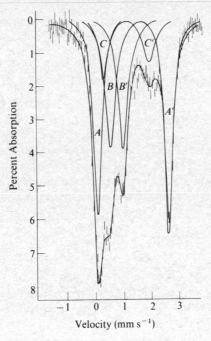

Fig. 7.3 Room temperature Mössbauer spectrum of deerite. [G. M. Bancroft, R. G. Burns, and A. J. Stone, *Geochim. Cosmochim. Acta*, **32**, 547 (1968)].

three doublets having centre shifts of 1.35 mm s^{-1} (A and A'), 1.12 mm s^{-1} (C and C') and 0.78 mm s^{-1} (B and B'). Some of these peaks are broad and indicate that they consist of two or more overlapping Lorentzians. The first two doublets have characteristic centre shifts for Fe^{2+} high spin and the latter doublet has a characteristic centre shift for Fe^{3+} high spin. The Fe^{2+}/Fe^{3+} ratio is approximately 2:1 as given by the chemical formula. Both the centre shift data and the area data strongly support the idea that there is little, if any, low spin iron. A small amount (less than about ten per cent of the total amount of iron) could not be easily detected because any low spin peak would strongly overlap peaks B and B'. The mechanism for how the magnetic moment is lowered is not at the present time well understood. Anti-ferromagnetic coupling is one possibility.

Although the assignment of oxidation state and electronic configuration of iron can normally be made unambiguously, there are several minerals to which the oxidation state of iron cannot readily be assigned from the Mössbauer centre shift. For example, pentlandite, $(Ni, Fe)_9S_8$, (Table 7.2) has two sets of lines with centre shifts of 0.83 and 0.62 mm s^{-1}, almost in between the 'normal' Fe^{2+} and Fe^{3+} high spin values. Similarly, wustite ($Fe_{\sim 0.9}O$) has two sets of lines with centre shifts of 1.02 and 1.14 mm s^{-1} (ref. 7.4). These latter values have been attributed to electron exchange, such that the oxidation state of the iron atoms is somewhere in between Fe^{2+} and Fe^{3+}.

7.2 Coordination number of iron in minerals

Table 7.1 and Fig. 7.1 summarize Mössbauer centre shifts for different known coordination numbers for ferrous and ferric species. It is apparent that the centre shifts generally increase with increasing coordination number, and these characteristic values can sometimes be useful in assigning the coordination number about iron in an unknown structure. Before looking at the uses of this correlation, several of the minerals in Table 7.1 will be briefly discussed because of their novel coordination polyhedra.

Almandine garnet has the general formula $(Mg, Fe)_3Al_2(SiO_4)_3$. The polyhedron about the (Mg, Fe) in this structure may be regarded as a distorted cube with four oxygen atoms at ~ 2.2 Å and four others at about 2.3 Å. This mineral gives the largest Fe^{2+} centre shift known in minerals. The six-coordinate structures include such common silicates as olivines, pyroxenes, and amphiboles. The latter two structures will be discussed in some detail in the next section. Staurolite has the idealized chemical formula $H_2Fe_4Al_{18}Si_8O_{48}$ and the structure contains chains of AlO_6 octahedra and FeO_4 tetrahedra sharing edges. Although some iron does enter the Al site(s), the majority is present in the tetrahedral site and gives rise to a significantly lower centre shift than six-coordinate minerals. Normal spinels have the composition AB_2O_4, A and B being divalent and trivalent ions respectively, and Fe^{2+} and Fe^{3+} normally give rise to similar centre shifts to appropriate four- or six-coordinate iron ions in silicates. In gillespite, Fe^{2+} ions occur in the rare square planar coordination. Fe–O distances are 1.95 Å to the nearest oxygens in the plane, but 3.98 Å to 4.75 Å to the nearest out of plane atoms. Gillespite gives an even lower centre shift than tetrahedrally coordinated Fe^{2+}.

Considering ferric species, epidote $[Ca_2(Al, Fe, Mn)_3Si_3O_{13}H]$ contains mainly Fe^{3+} in a very distorted six-coordinate site; whereas $FeCl_6^{\equiv}$, with

Fe^{3+} in a regular octahedral environment, gives a somewhat larger centre shift. Two reports of Mössbauer centre shifts of Fe^{3+} in four-coordination, in $FeCl_4^-$ and in the rare earth iron garnets $R_3Fe_2(FeO_4)_3$, show that four-coordinate Fe^{3+} has an appreciably smaller centre shift than six-coordinate Fe^{3+}. However, Fe^{3+} in orthoclase (Fe^{3+} substituting for Al in $KAlSi_3O_8$) shows an anomalously high value for four-coordinate ferric iron (ref. 7.5). Because of the very low amount of Fe^{3+} in the orthoclase specimen, and the resulting poor (and unshown) spectrum, it is difficult to assess the validity of this result.

It is apparent from the above centre shifts that there is a fairly small difference in centre shift between coordination numbers, making it sometimes difficult to assign coordination numbers conclusively.

Several examples illustrate the potential of Mössbauer spectroscopy for determining coordination numbers. The Mössbauer spectra of two sapphirines $(Mg, Fe^{2+})_2(Al, Fe^{3+})_4O_6SiO_4$ are shown in Fig. 7.4. These spectra give four distinct doublets with the following centre shifts: A and A', 1·41 mm s^{-1}; C and C', 1·39 mm s^{-1}; B and B', 0·56 mm s^{-1}; and D and D', 0·53 mm s^{-1}. These can readily be assigned (Fig. 7.1a): A and A', and C and C'

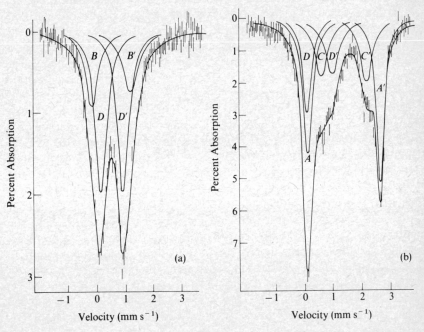

Fig. 7.4 Mössbauer spectra of two sapphirines: (a) yellow sapphirine; (b) blue-green sapphirine. [G. M. Bancroft, R. G. Burns, and A. J. Stone, *Geochim. Cosmochim. Acta*, **32**, 547 (1968).]

Mössbauer spectroscopy

to Fe^{2+} in two octahedral sites; B and B', and D and D' to Fe^{3+} in two tetrahedral positions. Although the centre shift for peaks D and D' in Fig. 7.4b is not very accurate because of the strong overlap of D with other peaks, the centre shifts for the Fe^{3+} species from Fig. 7.4a should be much more accurate. The above assignment appears to be reasonably convincing, and is consistent with an earlier suggestion that Fe^{2+} orders on six-coordinate sites, while Fe^{3+} orders on four-coordinate sites.

Another example concerns the assignment of Fe^{3+} to four- or six-coordinate sites in micas. A typical spectrum of a biotite is shown in Fig. 7.5.

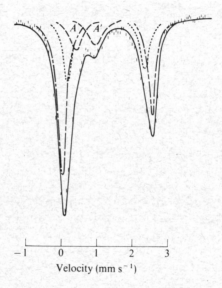

Fig. 7.5 Mössbauer spectrum of a biotite. [L. Häggstrom, R. Wappling, and H. Annersten, *Chem. Phys. Lett.*, **4**, 107 (1969).]

The outer doublets are due to Fe^{2+}, while peaks A and A' are due to Fe^{3+} iron. Different workers have obtained centre shifts for this ferric doublet from ~ 0.6 mm s^{-1} to ~ 0.8 mm s^{-1}, and in some cases, two ferric doublets have been 'resolved', one due to tetrahedral Fe^{3+} and one to octahedral Fe^{3+}. However, from the large range of reported centre shifts (due to the poor resolution of these peaks), it is apparent that such claims cannot be conclusive, especially since no χ^2 values have been published.

The centre shifts mentioned earlier for pentlandite (Table 7.2) have been used to assign Fe^{II} to octahedral and tetrahedral sites, since one centre shift is appreciably smaller (~ 0.2 mm s^{-1}) than the other centre shift. Without knowing that tetrahedral and octahedral sites were present in this structure, any assignment of coordination number from these results would at best be suggestive because of the electron exchange which apparently makes these values anomalously low for Fe^{II} high spin.

7.3 Determination of the number of distinct structural positions and the assignment of peaks in complex spectra

From the spectra shown in Section 3.3, and Figs 7.3, 7.4, and 7.5, it is apparent that many spectra consist of a number of very closely overlapping lines. Reasons for fitting a certain number of lines have been discussed in Section 3.3. In this section we will look more carefully at the relationship between structure and number of lines in a Mössbauer spectrum, and indicate how these lines can be assigned to the different structural positions.

Pyroxenes. The general chemical formula for pyroxenes can be expressed as XYZ_2O_6, where X refers to the M_2 position, Y refers to the M_1 position, and Z to the Si position. The M_2 position is normally occupied by Ca^{2+}, Mg^{2+}, Fe^{2+}, Mn^{2+} or Na^+, while the M_1 position is preferentially occupied by Mg^{2+}, Mn^{2+}, Al^{3+}, or Fe^{3+}, and the Si position sometimes has Al^{3+} substituted for Si. The pyroxenes can be classified into several groups, the most common being the ferromagnesium and calcic pyroxenes which can be represented by the trapezium in Fig. 7.6a. Closely related to the above pyroxenes are fassaite, in which there is substitution for Al in both Y and Z positions, and johannesite ($CaMnSi_2O_6$). Another well known series of pyroxenes can be represented by the triangle in Fig. 7.6b.

The characteristic pyroxene structure consists of infinite chains of SiO_3 groups linked by cations extending along the c axis. The tetrahedra fit with chains of octahedra, at least half of which contain the medium to small cations such as Mn^{2+}, Fe^{3+} or Fe^{2+}. An idealized pyroxene structure is viewed along the Z direction (Fig. 7.7). Substitution of ions of different sizes merely leads to a proportionate expansion or contraction of cell parameters with normally only small distortions of the pyroxene chains.

There are normally two crystallographically distinct positions labelled M_1 and M_2 (Fig. 7.7). The cations in the smaller M_1 positions are coordinated to six oxygen atoms in a nearly regular octahedron, while the cations in the larger M_2 positions are coordinated to a varying number of oxygens in a very distorted environment (Table 7.3). For example, the coordination number for the M_2 position varies from six to seven to eight in hypersthene, pigeonite and diopside respectively.

The majority of common pyroxenes have just one structural type of M_1 and M_2 position although the space group changes, for example, from Pbca for enstatite and hypersthene, to $P2_1/c$ for pigeonite, to C2/c for diopside and hedenbergite. However, for a few minerals such as spodumene [$LiAlSi_2O_6$] and some omphacites (Fig. 7.6), the space group symmetry is lowered to C2

Mössbauer spectroscopy

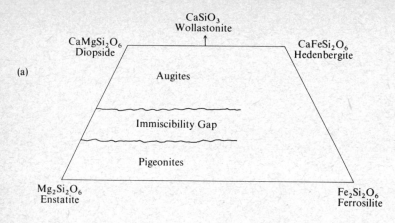

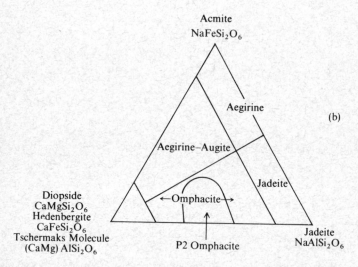

Fig. 7.6 Composition of important pyroxenes (a) in the system $CaMgSi_2O_6$–$Mg_2Si_2O_6$–$CaFeSi_2O_6$–$Fe_2Si_2O_6$; (b) in the system $NaFe^{3+}Si_2O_6$–$NaAlSi_2O_6$–$(CaMgFeAl)Si_2O_6$.

and $P2$ respectively. In spodumene, there are two M_1 and two M_2 positions, while in $P2$ omphacites there are four M_1 and four M_2 structurally distinct positions (ref. 7.6). Crystallographic data for omphacites suggests that the four M_2 positions are occupied by Na^+ and Ca^{2+} and the four M_1 positions by Mg, Fe^{2+}, Fe^{3+}, and Al. On the basis of bond distances and charge balance, it was postulated that Mg and Fe^{2+} occupy two of these M_1 positions and Fe^{3+} and Al occupy the other two, thus giving (Fe^{2+}, Mg) octahedra alternating with (Fe^{3+}, Al) octahedra.

Before discussing the expected Mössbauer spectra for the above pyroxenes, it should be noted that Fe^{2+} centre shifts (Table 7.1) are more sensitive to

Mössbauer spectroscopy as a fingerprint technique

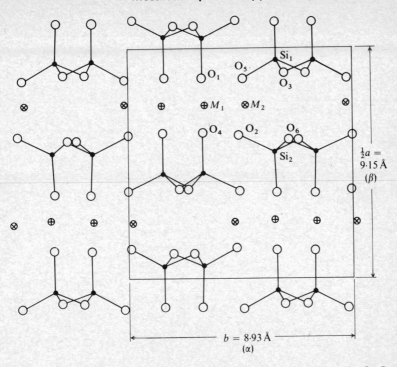

Fig. 7.7 The orthopyroxene crystal structure projected onto (001). (R. G. Burns, *Mineralogical Applications of Crystal Field Theory*, Cambridge University Press, 1970, Page 90).

coordination number changes than Fe^{3+} centre shifts. Similarly, the Fe^{2+} quadrupole splittings are much more sensitive than Fe^{3+} quadrupole splittings to changes in coordination number (Fig. 7.1b) and changes in site distortions. Thus, it is often possible to resolve Mössbauer peaks from Fe^{2+} in two or more positions, but it is almost always *not* possible to resolve Fe^{3+} peaks from two or more positions. Thus in pyroxenes, Fe^{3+} in the M_1 and M_2 positions results in a single doublet while Fe^{2+} in the M_1 and M_2 positions results in two doublets.† In pyroxenes with just one M_1 position and one M_2 position, we would expect a maximum of six lines from both Fe^{2+} and Fe^{3+}—two doublets from Fe^{2+} and one doublet from Fe^{3+}. For calcium and sodium end members such as $Ca(Fe, Mg) Si_2O_6$ or $Na(Al, Fe)Si_2O_6$ in which the larger Ca and Na ions exclusively occupy the larger M_2 positions, we would expect a maximum of two lines from Fe^{2+} or Fe^{3+}, and four lines if both Fe^{2+} and Fe^{3+} are present in solid solutions of these two end members. In contrast, as stated in section 3.3, for omphacites

† Recent spectra (ref. 7.7) of synthetic ferridiopside have resolved tetrahedral Fe^{3+} (substituting for Si) from Fe^{3+} in M_1 and M_2.

167

Table 7.3 Cation–oxygen bond lengths in pyroxenes*

Site		Hypersthene[1]	Ferrosilite[2]	Diopside[3]	Augite[3]	Pigeonite[4]	Omphacite[3]
M_1		2·087	2·206	2·115(2)	2·120(2)	2·158	2·095(2)
		2·036	2·093	2·065(2)	2·035(2)	2·062	2·032(2)
		2·075	2·200	2·050(2)	2·007(2)	2·162	1·994(2)
		2·038	2·119			2·068	
		2·152	2·101			2·057	
		2·166	2·152			2·086	
Mean		2·092	2·145	2·077	2·054	2·099	2·040
M_2		2·066	2·184	2·360(2)	2·316(2)	2·143	2·364(2)
		2·037	2·127	2·353(2)	2·278(2)	2·156	2·356(2)
		2·405	2·042	2·561(2)	2·563(2)	2·093	2·512(2)
		2·519	2·013	2·717(2)	2·760(2)	2·006	2·747(2)
		2·119	2·456			2·430	
		2·175	2·617			2·614	
						2·968	
Mean		2·220	2·240	2·425(6)	2·386(6)	2·240(6)	2·411(6)
				2·498(8)	2·479(8)	2·344(8)	2·495(8)

* No. of bonds are given in brackets.
[1] S. Ghose, *Z. Krist.*, **122**, 81 (1965).
[2] C. W. Burnham, *Ann. Repts. Geophys. Lab.*, **65**, 285 (1966).
[3] J. R. Clark, D. E. Appleman and J. J. Papike, *Mineral Soc. Amer. Spec. Pap.*, **2**, 31 (1969).
[4] N. Morimoto and N. Guven, *Ann. Repts. Geophys. Lab.* **66**, 494 (1968).

with eight possible positions for iron to enter, we might expect sixteen Fe^{2+} lines (eight doublets) and two Fe^{3+} lines (one doublet). From the sensitivity of Mössbauer parameters noted earlier, it is apparent that many of these lines are likely to overlap strongly.

The Mössbauer spectra of pyroxenes largely reflect the expected chemical and structural variations noted above. For example minerals along the diopside–hedenbergite tie line $Ca(Mg, Fe^{2+})Si_2O_6$ consist of two narrow (~ 0.3 mm s^{-1}) lines with a quadrupole splitting of about 2.0 mm s^{-1} (Table 7.4). This doublet can immediately be assigned to Fe^{2+} in the pyroxene M_1 position. Similarly, an aegirine-jadeite of composition $Na(Fe^{3+}, Al)Si_2O_6$ gives a two line spectrum having a centre shift of 0.65 mm s^{-1}—characteristic of Fe^{3+} in octahedral coordination (Table 7.4).

Table 7.4 Mössbauer parameters (mm s^{-1}) for important pyroxenes (at 295 K)

Mineral(s)	C.S.	Q.S.	Assignment
$Ca(Fe, Mg)Si_2O_6$	~ 1.45	1.9–2.3	Fe^{2+} in M_1
$Na(Al, Fe)Si_2O_6$	0.65	0.33	Fe^{3+} in M_1
$(Fe, Mg)_2Si_2O_6$	~ 1.4	1.9–2.1	Fe^{2+} in M_2
	~ 1.4	2.3–2.7	Fe^{2+} in M_1
$(Ca, Fe, Mg)_2Si_2O_6$	1.40	1.96	Fe^{2+} in M_2
	1.40	2.44	Fe^{2+} in M_1
	0.70	0.59	Fe^{3+} in (M_1, M_2)
$P2(Na, Ca, Al, Fe, Mg)_2Si_2O_6$	~ 1.4	~ 2.8	Fe^{2+} in one M_1
	~ 1.4	~ 2.3	Fe^{2+} in another M_1
	~ 1.4	~ 2.1	Fe^{2+} in another M_1
	~ 1.4	~ 1.8	Fe^{2+} in another M_1
	~ 0.7	~ 0.4	Fe^{3+} in M_1

G. M. Bancroft, P. G. L. Williams and R. G. Burns, *Amer. Mineral.*, **56**, 1617 (1971); G. M. Bancroft, P. G. L. Williams and E. J. Essene, *Min. Soc. Amer. Spec. Pap.*, **2**, 59 (1969), G. M. Bancroft, R. G. Burns and A. G. Maddock, *Geochim. Cosmochim. Acta*, **31**, 2219 (1967).

As suggested above, for orthopyroxenes $(Fe, Mg)_2Si_2O_6$, in which Fe^{2+} is present in both M_1 and M_2 positions, more than two lines are present. A spectrum of $(Fe_{0.86}Mg_{0.14})_2Si_2O_6$ shown in Figure 7.8 gives three visually resolved lines indicating the contribution of more than one type of iron atom to this spectrum. The two lines at high velocity must have components at low velocity to yield reasonable Fe^{2+} centre shifts. Thus this spectrum has been fitted to four lines due to two types of Fe^{2+}. Since Fe^{2+} centre shifts are normally quite similar, we pair these four lines off as follows: inner two lines (C.S. ~ 1.4 mm s^{-1}) to one type of Fe^{2+}, and outer two lines (C.S. ~ 1.4 mm s^{-1}) to another type of Fe^{2+}. Both of these values are characteristic

Mössbauer spectroscopy

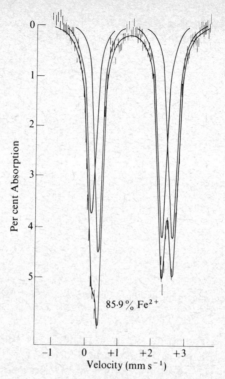

Fig. 7.8 Room temperature Mössbauer spectrum of an orthopyroxene $(Fe_{0.86}Mg_{0.14})_2Si_2O_6$. Outer peaks are due to Fe^{2+} in the M_1 positions; inner two peaks are due to Fe^{2+} in the M_2 position. (G. M. Bancroft, R. G. Burns, and R. A. Howie, *Nature*, **213**, 1221 (1967).)

of Fe^{2+} in six coordination. Any other pairing of these four peaks would give centre shifts for at least one doublet which are inconsistent with previous centre shifts from minerals and chemical compounds.

The unconstrained line widths for this spectrum are very similar, 0.29 ± 0.02 mm s^{-1}, but it is noticeable that the areas of component peaks of a doublet show small, but significant, differences. This spectrum shows no sign of the characteristic Fe^{3+} peaks, indicating that there is very little, if any, Fe^{3+} present in this sample.

To assign these two doublets to the M_1 and M_2 positions, it is necessary to look at spectra of other pyroxenes containing smaller amounts of iron. As the iron content decreases, the relative size of the outer two peaks decreases, but the room temperature resolution becomes poorer such that the peaks are not visually resolved. At liquid N_2 temperature however, the resolution improves as indicated by a spectrum of $(Fe_{1.10}Mg_{0.90})Si_2O_6$ (Fig. 7.9). Because component peaks of a doublet are normally of similar width and intensity, this spectrum confirms our previous assignment: outer

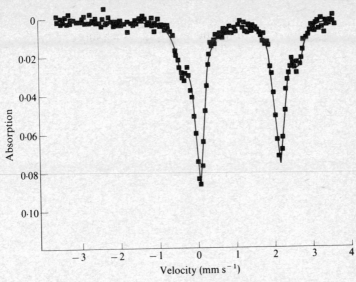

Fig. 7.9 Mössbauer spectrum of an orthopyroxene $(Fe^{2+}_{1.10}Mg_{0.90})Si_2O_6$ at 78 K. [D. Virgo and S. S. Hafner, *Min. Soc. Amer. Spec. Paper*, **2**, 67 (1969).]

peaks to one type of iron atom, and inner peaks to the other type of Fe atom. Because a previous X-ray structure indicated that Fe^{2+} preferred the M_2 position over the M_1 position, (chapter 8), it is reasonable to assign the orthopyroxene spectra as follows: inner peaks to Fe^{2+} in the M_2 position, and outer peaks to Fe^{2+} in the M_1 position. The smaller quadrupole splitting of the more distorted M_2 position is consistent with this assignment. As shown in section 6.5, the greater the distortion from octahedral symmetry, the smaller is the expected quadrupole splitting. The variations in quadrupole splitting with varying iron content will be discussed in section 7.6.

A spectrum of an augite (Fig. 7.10) demonstrates the six-line spectrum expected when both Fe^{2+} and Fe^{3+} are present in a C2/c pyroxene. Peaks 1 and 1' are due to Fe^{3+} in the M_1 and/or M_2 positions, while peaks C and C' are due to Fe^{2+} in M_2 and peaks A and A' are due to Fe^{2+} in M_1 as in the orthopyroxenes. The parameters are given in Table 7.4, and it is apparent that the Fe^{2+} parameters for both M_1 and M_2 positions are very similar to those for orthopyroxenes, but that Fe^{2+} in the augite M_1 position gives a significantly larger quadrupole splitting than Fe^{2+} in the diopside–hedenbergite M_1 position. As for orthopyroxenes, the spectra of this augite at 77 K and 295 K give almost identical relative areas.

There are two features of both the room and low temperature spectra of this augite that deserve comment. First, unlike the orthopyroxene peaks, the line widths for this augite are very broad (~ 0.45 mm s^{-1}) for lines known to arise from one type of iron atom. This broadening will be discussed in

Mössbauer spectroscopy

section 7.4. Second, the areas of the component ferrous peaks are not equal. In particular, peak C is much more intense than C', and A less intense than A'. These inequalities mostly reflect the poor resolution and resulting large

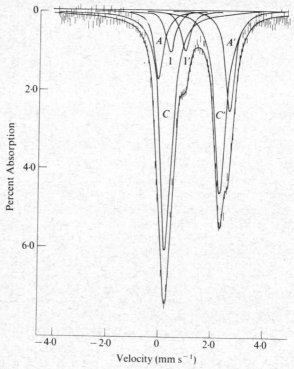

Fig. 7.10 Room temperature Mössbauer spectrum of an augite containing 0·61 Ca^{2+} and 0·20 Fe^{2+} per formula unit. [P. G. L. Williams, G. M. Bancroft, M. G. Bown, and A. C. Turnock, *Nature*, **230**, 149 (1971).]

errors in the areas of A and C, but part of the inequalities may be due to other effects (section 7.4).

It is apparent from the quadrupole splittings in Table 7.4, that there is a very large variation in pyroxene M_1 quadrupole splittings at room temperature (and also at lower temperatures). For example, diopside $CaMg(Fe)Si_2O_6$ gives an M_1 quadrupole splitting of 1·89 mm s^{-1}, while some orthopyroxenes, and the outer lines of omphacites have quadrupole splittings of about 2·8 mm s^{-1}. These M_1 quadrupole splittings are normally temperature dependent. For example, a typical increase in the M_1 quadrupole splitting from 295 K to 78 K is 0·5 mm s^{-1}, although this increase varies somewhat (± 0.2 mm s^{-1}) from pyroxene to pyroxene. In contrast, the M_2 quadrupole splitting has much more constant values for all pyroxenes (2·0 $\pm$ 0·1 mm s^{-1}), and this quadrupole splitting is normally much more temperature independent. A typical increase is 0.1 mm s^{-1} from 295 K to 78 K. This temperature dependence will be discussed again in section 7.6.

The above M_1 and M_2 quadrupole splittings are helpful in assigning the very complex spectra of omphacites (Fig. 3.7). We have already discussed (chapter 3) the reasons for fitting ten peaks (four Fe^{2+} doublets, and one Fe^{3+} doublet) to these spectra. The assignment of the four ferrous doublets presents considerable difficulties. There appears to be two tenable assignments: first, all four doublets arise from Fe^{2+} in the four M_1 positions; and second, two doublets (the inner two) are due to Fe^{2+} in two or four M_2 positions, and the outer two doublets are due to Fe^{2+} in two or four M_1 positions. The first assignment seems to be more likely for the following reasons. First, despite the small quadrupole splitting of the inner doublets, all four doublets are relatively temperature dependent—the quadrupole splitting increases by about 0.4 mm s^{-1} from 295 K to 77 K. This increase is similar to that for Fe^{2+} in other pyroxene positions. The smallest quadrupole splitting (~ 1.8 mm s^{-1}) is similar to that found in diopside. Second, from the very small Ca discrepancies in the M_2 position determined from chemical analyses (~ 0.01 per formula unit), it seems unlikely that about half of the ~ 0.06 Fe^{2+} per formula unit (from the Mössbauer areas) enters the M_2 positions. If ~ 0.03 Fe^{2+} per formula unit did enter the M_2 positions, this would imply that either the Na + Ca chemical analysis has a large error, or that Na and/or Ca enters the M_1 sites. Both of these seem unlikely, and taken together with the temperature dependencies of the quadrupole splittings, the assignment of the four doublets to the four M_1 positions is strongly suggested.

In an attempt to assign the M_1 peaks to specific positions in the structure, it should be noted that Clark and Papike's X-ray data indicate that Fe^{2+} orders in the M_1 and M_1 (1) H positions, while the small amount of Fe^{3+} enters (along with Al) the M_1 (1) and M_1H positions. The Mössbauer spectra indicate that appreciable amounts of Fe^{2+} enter all four positions. Peaks 1 and 3 (Fig. 3.7) are generally more intense than 2 and 4, and using the crystallographic evidence, we tentatively assign peaks 1 and 3 to the M_1 and M_1 (1) H positions (not necessarily respectively). Peaks 2 and 4 can then be assigned to the M_1 (1) and M_1H positions. The one ferric doublet is assigned to any or all of the M_1 positions.

Most of the above spectra gave no indication of impurities. Generally, (except for the augite) narrow (<0.35 mm s^{-1}) lines were observed, and more importantly, the spectra at a range of temperatures were consistent, i.e. the relative areas of peaks remained constant over a large temperature range. However, some omphacite spectra give a peak at about 1.6 mm s^{-1} which is indicative of an epidote impurity. This peak lies in between the two main absorption areas in omphacites and so can immediately be recognized. However, there are other examples, where impurities and/or inhomogeneous (perhaps on a micro level) compositions have given rise to extreme difficulties

Mössbauer spectroscopy

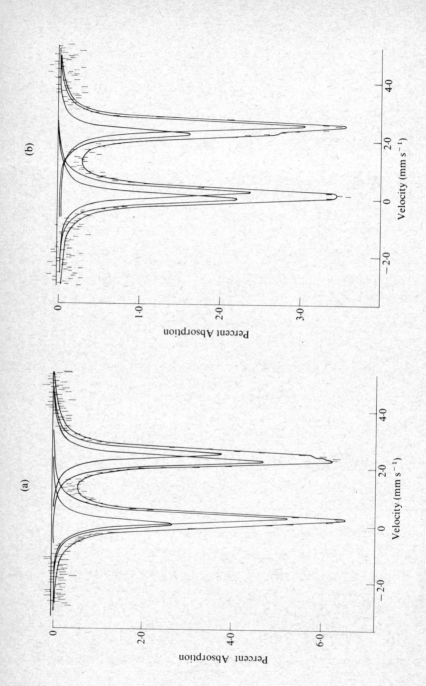

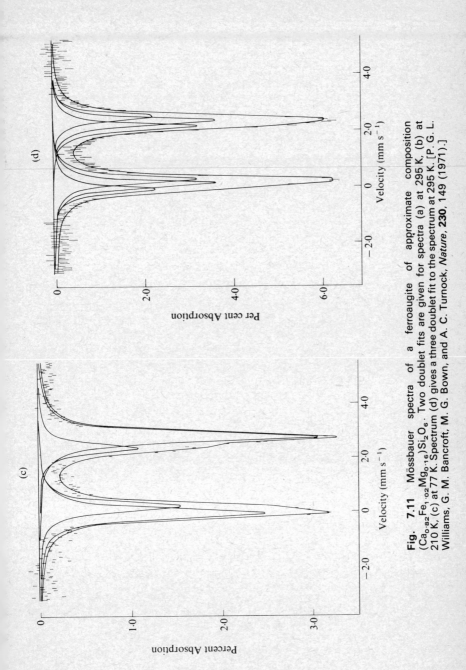

Fig. 7.11 Mössbauer spectra of a ferroaugite of approximate composition $(Ca_{0.82}Fe_{1.02}Mg_{0.16})Si_2O_6$. Two doublet fits are given for spectra (a) at 295 K, (b) at 210 K, (c) at 77 K. Spectrum (d) gives a three doublet fit to the spectrum at 295 K. [P. G. L. Williams, G. M. Bancroft, M. G. Bown, and A. C. Turnock, *Nature*, **230**, 149 (1971).]

Mössbauer spectroscopy

in interpretation and the following two examples point out the necessity of using other techniques such as X-ray diffraction, electron probe, and electron microscope in conjunction with Mössbauer. An original spectrum of a hedenbergite of approximate composition $Ca_{0.95}Fe_{1.05}Si_2O_6$ at room temperature gave an intense inner doublet with a quadrupole splitting of ~ 2.2 mm s^{-1}, and a very weak outer doublet of quadrupole splitting ~ 2.8 mm s^{-1}. From the above composition, it seemed obvious that the inner doublet was due to Fe^{2+} in M_1, and the outer doublet was due to Fe^{2+} filling the Ca discrepancy in the M_2 position. However, it soon became apparent that the above M_2 quadrupole splitting was anomalously large (Table 7.4.). A more detailed study at a variety of temperatures revealed another doublet with a quadrupole splitting of ~ 2.0 mm s^{-1} at 295 K, and this is more likely due to Fe^{2+} in M_2 (Table 7.4). The outer doublet (Q.S. ~ 2.8 mm s^{-1}) is more likely due to a very small amount of coexisting actinolite, the M_1 peaks of which have a quadrupole splitting of ~ 2.8 mm s^{-1} (Table 7.7).

The spectra of a large number of synthetic and natural augites (Fig. 7.11) have presented a number of problems. At room temperature, minerals of approximate composition $(Ca_{0.8}Fe_{0.8}Mg_{0.4})Si_2O_6$ give spectra showing an intense inner doublet and a weaker outer doublet as in Fig. 7.11a with $A_{inner}/A_{outer} \sim 2:1$. If we assign this spectrum as previously—inner doublet to Fe^{2+} in M_2 and outer doublet to Fe^{2+} in M_1—the above area ratio (chapter 8) indicates that over 0.5 Fe^{2+} per formula unit enters M_2 implying that ~ 0.3 Ca^{2+} per formula unit enters the M_1 position, and this appears to be crystal chemically untenable. Because of this difficulty, the reverse assignment was proposed; outer doublet due to Fe^{2+} in M_2, and inner doublet to M_1. However, this assignment gives anomalously large M_2 quadrupole splittings of ~ 2.5 mm s^{-1}. The 77 K spectrum (Fig. 7.11c) shows the reverse effect: strong outer doublet and weaker inner doublet with $A_{inner}/A_{outer} < 0.5$. In addition, the inner doublet had a temperature-independent quadrupole splitting, strongly indicating that the *inner* doublet was the M_2 doublet. These spectra indicate strongly that there are more than two doublets present, with different temperature dependencies such that two (or more) doublets strongly overlap at 295 K to give an intense inner line, while one doublet shifts from 295 K to 77 K to overlap strongly with the outer doublet. (Fig. 7.11d).

Detailed examination by other techniques indicated no appreciable non-C2/c phases present. There appears to be two tenable explanations for the above phenomena. First, it is possible that there are two or more exsolved C2/c phases present on a very fine scale, which give significantly different quadrupole splittings. Second, it has been recently suggested that the chemical disorder of Ca^{2+} and (Fe^{2+}, Mg^{2+}) in the M_2 site gives four next-nearest neighbour configurations for iron cations in the M_1 site, and each one of these four configurations gives a distinct doublet. It is difficult to resolve these four M_1 doublets. This kind of effect may also be important in the complex omphacite spectra, but usually this type of chemical disorder just gives line broadenings (section 7.4).

Amphiboles. The chemical formula for amphiboles can be expressed generally as $X_2Y_5Z_8O_{22}(OH)_2$, where X refers to the M_4 position, Y refers the M_1, M_2, and M_3 positions, and Z to the Si position. The M_4 position, like the M_2 position in pyroxenes, is occupied preferentially by the larger cations such as Ca^{2+}, Na^+ or Fe^{2+}, while the M_1, M_2, and M_3 positions (denoted M_{123}) are preferentially occupied by the smaller cations such as Mg^{2+}, Fe^{3+} and Al^{3+}. Once again, Si is sometimes replaced by Al. In addition to these four positions, another position, the so-called A position, is sometimes occupied by any excess Na or K above that given by the chemical formula.

Table 7.5 Idealized cation distribution in some of the important magnesium and ferrous-end members of the amphiboles $X_2Y_5Z_8O_{22}(OH)_2$

Mineral	X	Y	Z
1. Cummingtonite (anthophyllite)	Mg_2	Mg_5	Si_8
2. Grunerite	Fe_2^{2+}	Fe_5^{2+}	Si_8
3. Gedrite	Mg_2	Mg_3Al_2	Si_6Al_2
4. Ferrogedrite	Fe_2^{2+}	$Fe_3^{2+}Al_2$	Si_6Al_2
5. Tremolite	Ca_2	Mg_5	Si_8
6. Ferroactinolite	Ca_2	Fe_5^{2+}	Si_8
7. Hornblende	Ca_2	Mg_4Al	Si_7Al
8. Tschermakite	Ca_2	Mg_3Al_2	Si_6Al_2
9. Ferrotschermakite	Ca_2	$Fe_3^{2+}Al_2$	Si_6Al_2
10. Glaucophane	Na_2	Mg_3Al_2	Si_8
11. Ferroglaucophane	Na_2	$Fe_3^{2+}Al_2$	Si_8
12. Riebeckite	Na_2	$Fe_3^{2+}Fe_2^{3+}$	Si_8
13. Magnesioriebeckite	Na_2	$Mg_3Fe_2^{3+}$	Si_8

Some important amphibole end members are given in Table 7.5. The substitution of Mg^{2+} for Fe^{2+} and vice versa involves no charge unbalance, but is of prime importance in consideration of many physical properties of amphiboles. Four other major substitutions are important in amphiboles, the first three of which are indicated by Table 7.5; i.e. Al $\rightleftarrows$ Si, (Mg, Fe) $\rightleftarrows$ Al, Na $\rightleftarrows$ Ca and the introduction of Na into the A site. As in pyroxenes, solid solution series between the end members are very common.

Mössbauer spectroscopy

The basic unit of all amphibole structures is (Si, Al)O_4 tetrahedra linked to form chains of composition $(Si_4O_{11})_n$ which have double the width of those in pyroxenes (Fig. 7.12). These Si_4O_{11} groups are linked by bands of cations that extend along the c axis. A projection of the cummingtonite structure onto (001) is shown in the next figure (Fig. 7.13). The four major cation

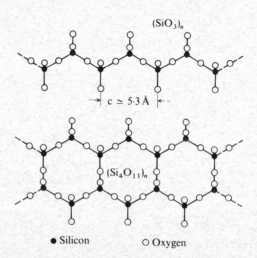

Fig. 7.12 Comparison of amphibole band $(Si_4O_{11})_n$ and pyroxene chain $(SiO_3)_n$. After W. F. de Jong, 1959, *General Crystallography*, Freeman, San Francisco, and W. A. Deer, R. A. Howie and J. Zussman, *An Introduction to the Rock Forming Minerals*, Longmans (1967).

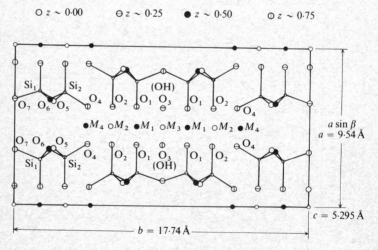

Fig. 7.13 Crystal structure of an amphibole projected onto (001). From G. M. Bancroft and R. G. Burns, *Min. Soc. Amer. Spec. Paper*, **2**, 137 (1969).

Table 7.6 Cation–oxygen bond lengths in amphiboles

Atoms	Bond multiplicity	Tremolite[1]	Actinolite[2]	Cummingtonite[3]	Grunerite[4]	Glaucophane[1]	Hornblende[1]
$M_1 - O_1$	2	2·064		2·076	2·083	2·078	2·042
$M_1 - O_2$	2	2·078		2·115	2·164	2·082	2·141
$M_1 - O_3$	2	2·083		2·103	2·131	2·100	2·043
mean		2·075	2·104	2·098	2·126	2·087	2·075
$M_2 - O_1$	2	2·133		2·128	2·162	2·038	2·101
$M_2 - O_2$	2	2·083		2·093	2·130	1·943	2·072
$M_2 - O_4$	2	2·014		2·029	2·078	1·849	1·973
mean		2·077	2·075	2·083	2·123	1·943	2·049
$M_3 - O_1$	4	2·070		2·102	2·128	2·103	2·083
$M_3 - O_3$	2	2·057		2·070	2·110	2·077	2·076
mean		2·066	2·097	2·091	2·122	2·094	2·081
$M_4 - O_2$	2	2·397		2·176	2·144	2·411	2·394
$M_4 - O_4$	2	2·321		2·041	1·988	2·337	2·328
$M_4 - O_6$	2	2·539		2·699	2·764	2·446	2·582
$M_4 - O_5$	2	2·767		3·147	3·290	2·798	2·665
mean(6)		2·419		2·305	2·299	2·398	2·435
mean(8)		2·506		2·516	2·546	2·498	2·492

[1] J. J. Papike, M. Ross and J. R. Clark, *Mineral. Soc. Amer. Spec. Pap.*, **2**, 117 (1969).
[2] J. T. Judson, F. D. Bloss and G. V. Gibbs, *Amer. Mineral.*, **55**, 302 (1970).
[3] K. F. Fischer, *Amer. Mineral.*, **51**, 814 (1966).
[4] L. W. Finger and T. Zoltai, *Trans. Amer. Geophys. Union*, **48**, 233 (1967).

positions are in the ratio $M_4:M_3:M_2:M_1 = 2:1:2:2$. Cations in the M_1 and M_3 positions are each coordinated in an almost regular octahedron to four oxygens and two oxygens of hydroxyl groups, whereas six oxygens surround the M_2 cations in a slightly more distorted site (Table 7.6). The very distorted M_4 site is much larger than the other three, and is variously described as six or eight coordinate.

By analogy with the pyroxene M_2 position, the larger cations such as Ca^{2+} or Na^+ preferentially enter the M_4 position, and thus in an alkali amphibole such as riebeckite, $Na_2Fe_3^{2+}Fe_2^{3+}Si_8O_{22}(OH)_2$ we might expect a maximum of eight lines in the Mössbauer spectrum: three doublets from Fe^{2+} in the M_1, M_2, and M_3 positions, and one doublet from Fe^{3+} in any or all of these three positions. For a ferromagnesium amphibole such as grunerite $Fe_7^{2+}Si_8O_{22}(OH)_2$, we might expect a maximum of eight lines from the Fe^{2+} in the four positions, giving rise to a very complex spectrum similar to those of omphacites.

Coincidentally, the ferromagnesium cummingtonites, anthophyllites, and grunerites† give the simplest amphibole spectra (Figs 3.6 and 7.14). They consist of two relatively narrow doublets, with the inner two peaks having widths of ~ 0.30 mm s^{-1}, and the outer two peaks having slightly broader lines (~ 0.35 mm s^{-1}) suggesting that each outer peak is a superposition of at least two lines. The resolution of the two doublets increases on lowering the temperature, but no further fine structure becomes noticeable. As with the orthopyroxenes, the peaks are again assigned: inner two peaks to one type of Fe^{2+}, and outer two peaks to another type of Fe^{2+}. This assignment gives reasonable Fe^{2+} centre shifts for both doublets (Table 7.7), and component peaks of a doublet are of similar area, although the low velocity components are normally slightly more intense than their corresponding high velocity counterparts. An X-ray structure of a cummingtonite of approximate composition $(Fe_{0.35}Mg_{0.65})_7Si_8O_{22}(OH)_2$ showed that Fe^{2+} strongly preferred the M_4 position. The spectrum of a cummingtonite of similar composition (Fig. 7.14a) shows that the inner two peaks are the most intense, and it is thus reasonable to assign the spectrum as follows: inner doublet to Fe^{2+} in M_4, and outer doublet to a coincidental overlap of Fe^{2+} in M_1, M_2, and M_3. As for the orthopyroxene M_2 position, the smaller quadrupole splitting for the more distorted M_4 position is consistent with this assignment. Also, the larger widths of the outer two peaks strongly suggests that they are a superposition of lines with slightly different positions. The quantitative area ratios for these spectra (chapter 8) provide additional evidence for this assignment.

† The monoclinic cummingtonite–grunerite series extends from minerals containing about 30% of the $Fe_7Si_8O_{22}(OH)_2$ component to the grunerite end member. Cummingtonite refers to a range of minerals from 30% to 70% $Fe_7Si_8O_{22}(OH)_2$, with grunerite being used for more iron rich minerals. The magnesium rich anthophyllites are orthorhombic and contain up to 40% $Fe_7Si_8O_{22}(OH)_2$.

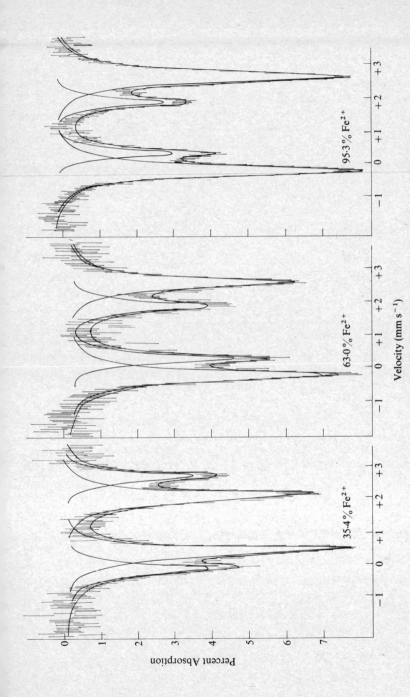

Fig. 7.14 Room temperature Mössbauer spectra of a cummingtonite and two grunerites [G. M. Bancroft, R. G. Burns and A. G. Maddock, *Amer. Mineral.*, **52**, 1009 (1967).]

Mössbauer spectroscopy

Table 7.7 Mössbauer parameters (mm s^{-1}) for important amphiboles (at 295 K)

Mineral(s)	C.S.	Q.S.	Assignment
(Fe, Mg)$_7$Si$_8$O$_{22}$(OH)$_2$ (cummingtonite-	~1·3	1·5–1·7	Fe^{2+} in M_4
grunerite)	~1·4	2·7–2·9	Fe^{2+} in M_{123}
(Fe, Mg)$_7$Si$_8$O$_{22}$(OH)$_2$ (anthophyllites)	1·35	1·80	Fe^{2+} in M_4
	1·37	2·6	Fe^{2+} in M_{123}
Ca$_2$(Fe, Mg)$_5$Si$_8$O$_{22}$(OH)$_2$	~1·4	2·89	Fe^{2+} in M_1
	~1·4	~1·9	Fe^{2+} in M_2
	~1·4	~2·4	Fe^{2+} in M_3
Na$_2$(Fe^{3+}, Fe^{2+}, Mg, Al)Si$_8$O$_{22}$(OH)$_2$	~1·4	~2·8	Fe^{2+} in M_1
	1·39	2·00	Fe^{2+} in M_2
	1·36	2·41	Fe^{2+} in M_3
	~0·65	~0·45	Fe^{3+} in M_{123}

G. M. Bancroft, A. G. Maddock and R. G. Burns, *Geochim. Cosmochim. Acta*, **31**, 2219 (1967); G. M. Bancroft and R. G. Burns, *Mineral. Soc. Amer. Spec. Pap.*, **2**, 137 (1969); R. G. Burns and C. Greaves, *Amer. Mineral.*, **56**, 2010 (1971).

The actinolite, alkali amphibole, and hornblende spectra are conveniently discussed together as their spectra are qualitatively similar. The M_4 position in these structures is occupied by Ca (actinolite and hornblende) or Na (alkali amphiboles), and as discussed earlier, we might expect a maximum of eight peaks in a sample containing both Fe^{2+} and Fe^{3+}. The spectra of two alkali amphiboles of approximate composition Na$_2$Fe$_3^{2+}$Fe$_2^{3+}$Si$_8$O$_{22}$(OH)$_2$ and Na$_2$(Mg$_{2·5}$Fe$_{1·0}^{2+}$Fe$_{1·5}^{3+}$)Si$_8$O$_{22}$(OH)$_2$ are shown in Figs. 7.15a and 7.15b respectively. In Fig. 7.15a, six peaks have been fit to the spectrum and, using arguments outlined for previous pyroxenes and amphiboles, we can assign the peaks as follows: peaks B and B' to Fe^{3+}, and peaks A and A', and C and C' to two different types of Fe^{2+}. Peaks C and C' are always broader than A and A', strongly suggesting that peaks C and C' are due to a superposition of two or more peaks. In Fig. 7.15b, resolution of the spectrum into four doublets has been attained, and the parameters for this four doublet fit (Table 7.7) are very similar to the corresponding four doublet fits to actinolites.

The ferric peaks can immediately be assigned to any, or all, of the M_{123} sites, although other spectral and crystallographic evidence strongly suggests that most of the ferric enters the M_2 position. In some actinolites, very broad ferric peaks are obtained, indicating that these peaks are due to a superposition of lines from Fe^{3+} in M_1 and/or M_3, as well as M_2.

The assignment of the ferrous peaks presents considerable difficulties, and other spectral and crystallographic data is essential to assign these peaks with confidence. Infrared data on the crocidolite, whose Mössbauer spectrum

Mössbauer spectroscopy as a fingerprint technique

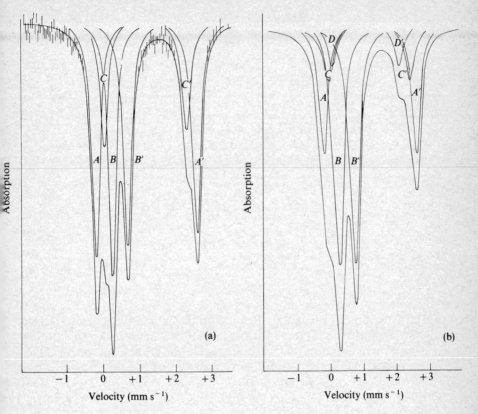

Fig. 7.15 Room temperature Mössbauer spectra of two alkali amphiboles, (a) ferrous crocidolite, (b) magnesioriebeckite. [G. M. Bancroft and R. G. Burns, *Min. Soc. Amer. Spec. Paper*, **2**, 137 (1969).]

is given in Fig. 7.15a, indicated the following approximate cation distribution for Fe^{2+} (per formula unit): $Fe^{2+}_{(M_1+M_3)} = 2.53$, $Fe^{2+}_{M_2} = 0.23$, with the amount of Fe^{2+} per formula unit from chemical analysis being 2.76. In addition, it was noted that there was more Fe^{2+} in M_1 than M_3. From the Mössbauer spectrum (Fig. 7.15a), peaks A and A' are about twice the intensity of C and C', i.e. $A_A/A_C \sim 2$, and thus $Fe^{2+}_A \sim 1.8$ and $Fe^{2+}_C \sim 0.9$. Combining the Mössbauer and infrared data, the only possible assignment of these peaks is: A and A' to Fe^{2+} in M_1, and C and C' to Fe^{2+} in $M_2 + M_3$. This assignment is consistent with the greater width of lines C and C'.

For sample b (Fig. 7.15b) and the actinolites, the three ferrous doublets may be assigned as: A and A' to Fe^{2+} in M_1 as above, C and C' to Fe^{2+} in M_3, and D and D' to Fe^{2+} in M_2. The M_2 site, being more distorted from

Mössbauer spectroscopy

octahedral symmetry than the M_1 or M_3 (Table 7.6) would be expected to give a smaller quadrupole splitting. The site populations for some of these minerals using this assignment are reasonably consistent with those from X-ray and infrared work. For example, since infrared and X-ray data strongly indicates that most of the 1·5 Fe^{3+} per formula unit enters the smaller M_2 position, we would expect a fairly small amount of Fe^{2+} in the available remaining M_2 positions. As the Mössbauer spectrum shows, only a comparatively small amount of Fe^{2+} enters M_2.

The differences between the cummingtonite–grunerite spectra and the above spectra are striking, and perhaps surprising considering the comparatively small differences in Fe–O bond lengths (Table 7.6). The M_2 quadrupole splitting has decreased from $\sim 2\cdot 8$ mm s^{-1} in the ferromagnesium amphiboles to $\sim 2\cdot 0$ mm s^{-1}, while the M_3 quadrupole splitting has decreased from $\sim 2\cdot 8$ to $2\cdot 4$ mm s^{-1}.

The spectra of hornblendes are similar in form to the above actinolite and alkali amphibole spectra, but we would hope to detect Fe^{3+} replacing Si or Al in the tetrahedral site. Although the Fe^{3+} peaks often broaden somewhat, it is very difficult to obtain accurate centre shifts and quadrupole splittings for more than one Fe^{3+} doublet.

Micas. The general chemical formula for micas can be expressed as $X_2Y_{4-6}Z_8O_{20}(OH, F)_4$: where X is mainly large cations such as K^+, Na^+, or Ca^{2+}; Y is mainly Al^{3+}, Mg^{2+}, or Fe^{2+}; and Z is mainly Si or Al^{3+}, with perhaps some Fe^{3+}. The micas can be divided into dioctahedral and trioctahedral classes (Table 7.8), in which the number of Y ions is 4 and 6

Table 7.8 Approximate chemical formulae of some micas, $X_2Y_{4-6}Z_8O_{20}(OH, F)_4$

	Dioctahedral		
	X	Y	Z
Muscovite	K_2	Al_4	Si_6Al_2
Paragonite	Na_2	Al_4	Si_6Al_2
Glauconite	$(K, Na)_{1\cdot2-2\cdot0}$	$(Fe, Mg, Al)_4$	$Si_{7-7\cdot6}Al_{1\cdot0-0\cdot4}$
	Trioctahedral		
	X	Y	Z
Phlogopite	K_2	$(Mg, Fe^{2+})_6$	Si_6Al_2
Biotite	K_2	$(Mg, Fe, Al)_6$	$Si_{6-5}Al_{2-3}$
Zinnwaldite	K_2	$(Fe, Li, Al)_6$	$Si_{6-5}Al_{2-3}$

(W. A. Deer, R. A. Howie, and J. Zussman, *An Introduction to the Rock Forming Minerals*, Longmans (1967)).

respectively. A summary of the approximate chemical formula of important micas is given in Table 7.8.

The basic structural feature of a mica is a composite sheet in which a layer of octahedrally coordinated cations (Y) is sandwiched between two identical layers of linked (Si, Al)O_4 tetrahedra (Fig. 7.16). The net negative charge of the (Si, Al)O_4 layers is balanced by planes of X ions (K, Na, etc.)

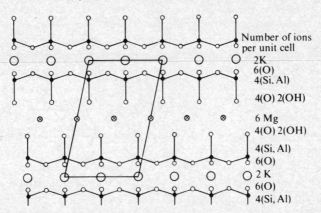

Fig. 7.16 Mica structure [from W. A. Deer, R. A. Howie, and J. Zussman, *An Introduction to the Rock Forming Minerals*, Longmans (1967)].

lying between them, and the repeat distance perpendicular to the sheets is approximately 10 Å or a multiple of 10 Å. The X ions are in twelve coordination. There are six different ways of stacking one sheet relative to its neighbour, and various sequences of layer rotations are possible such that when repeated regularly, these build up unit cells with 1, 2, 3 or more layers.

The crystal structure of muscovite has been determined, and there are two different octahedral sites for the Y ions, distinguishable by both size and symmetry. One site, generally occupied by Al, has an average cation–anion distance of 1·95 Å. The other site–vacant in the idealized dioctahedral formula—has an average cation–anion distance of 2·20 Å. The ratio of small sites to large sites is two to one.

The structure of phlogopite (a trioctahedral mica) is qualitatively similar to that of muscovite except that all the octahedral cation–anion distances are virtually the same, the average cation–anion distance being $\sim 2\cdot 11$ Å for both positions. The structure of zinnwaldite is less accurately known than the above two structures, but there appears to be two octahedral sites as in muscovite.

The above structural data indicate that a maximum of two ferrous doublets would arise in the Mössbauer spectra of these samples, with perhaps a broad Fe^{3+} doublet due to Fe^{3+} in one or both octahedral sites and perhaps in the Si site.

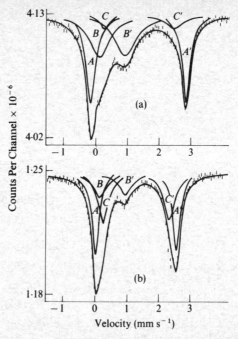

Fig. 7.17 Room temperature Mössbauer spectra of micas: (a) dioctahedral muscovite and (b) trioctahedral biotite. [C. S. Hogg and R. E. Meads, *Min. Mag.*, **37**, 606 (1970).]

The spectra (Fig. 7.17) are consistent with these structural data, but there are considerable ambiguities in assigning the peaks. The spectra are similar in form to the amphibole spectra, and from centre shifts (Table 7.9) we can assign the peaks: A and A', and C and C' to Fe^{2+} and B and B' to Fe^{3+}. In the muscovite spectrum (Fig. 7.17a) doublet A and A' was assigned to the larger site, and C and C' to the smaller site, indicating that Fe^{2+} strongly prefers the large site. This assignment is consistent with the suggestion of octahedral ordering proposed earlier on the basis of bond lengths, i.e., that small trivalent ions such as Al^{3+} and Fe^{3+} occupy the smaller site, and larger ions such as Mg^{2+} and Fe^{2+}, the larger site. The ferric peaks are broad, and suggest that Fe^{3+} enters two or more positions in the muscovite structure.

The spectra of trioctahedral micas (Fig. 7.17b) are very similar to those of muscovites. The outer two lines, however, have an appreciably lower quadrupole splitting (Table 7.9), and indicate that Mössbauer spectra should be very useful in distinguishing between dioctahedral and trioctahedral micas. For example, in Table 7.9, the outer quadrupole splitting for dioctahedral muscovites is greater than 3·0 mm s^{-1}, while the trioctahedral samples give quadrupole splittings generally less than 2·6 mm s^{-1}. The assignment of the peaks for the trioctahedral micas is very difficult, and different workers have

Table 7.9 Mössbauer parameters for sheet silicates (mm s^{-1} at 295 K)

Sample no.	% FeO	Fe^{2+} Outer		Fe^{2+} Inner		Fe^{3+}	
		C.S.	Q.S.	C.S.	Q.S.	C.S.	Q.S.
1	1·12	1·41	3·02	1·39	2·25	0·60	0·82
2	0·91	1·41	3·04	1·36	2·14	0·63	0·74
3	34·6	1·38	2·56	1·36	2·18	0·80	0·52
4	42	1·40	2·56	1·36	2·17	0·79	0·55
5	16·35	1·33	2·58	1·32	2·14	0·65	0·72
6	16·95	1·39	2·74	1·39	2·44	0·64	0·83
7	3·21	1·36	2·56	1·34	2·24	0·73	0·80
8	5·70	1·41	2·65			0·66	0·89

Samples
1. Lee Moor muscovite (Fig. 7.16a)
2. Madagascar muscovite
3. Biotite
4. Synthetic annite
5. Aberdeen biotite (Fig. 7.16b)
6. Trelavour biotite
7. Indian phlogopite
8. Zinnwaldite

L. Häggstrom, R. Wappling and H. Annersten, *Chem. Phys. Lett.*, **4**, 107 (1969); C. S. Hogg and R. E. Meads, *Min. Mag.*, **37**, 606 (1970).

assigned the two doublets in the two possible ways. Also the poorer resolution of the trioctahedral mica spectra makes it impossible to determine accurately the areas of the peaks. (chapter 8). It is possible that better resolution of mica spectra could be attained at other than room temperature.

Other applications to weathering and oxidation processes in clay minerals will be further discussed in section 6 of this chapter.

Epidotes. Although many other spectra could be described which illustrate the relationship between Mössbauer spectra and structure, a recent spectrum of allanite $Ca_{1·00}(RE_{0·74}Ca_{0·26})$ $(Al_{0·66}Fe_{0·34}Al_{1·00})(Al_{0·17}Fe_{0·83})Si_3O_{13}H$ (Fig. 7.18) combined with detailed structural data, provides a fine example. Allanite belongs to the epidote groups of minerals, which, as we have seen previously, has the basic chemical formula $Ca_2(Al, Fe^{3+}, Mn^{3+})_3Si_3O_{13}H$. In allanite, trivalent rare earth (RE) ions substitute for Ca, with divalent ions such as Fe^{2+} substituting for Al to make up the charge balance.

Epidotes have three cation sites, M_1, M_2 and M_3 into which Fe^{2+}, Fe^{3+} and Al^{3+} enter. The M_3 (Al, Fe)O_6 polyhedron is very irregular while the M_1 and M_2 sites are more regularly octahedral, but vary considerably in different members of the epidote group. Previous evidence—optical properties and absorption spectra—showed that in the Al–Fe epidotes, Fe^{3+} occupies

Mössbauer spectroscopy

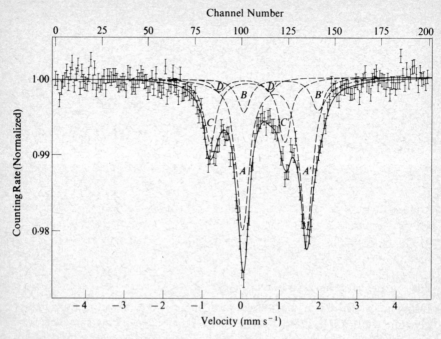

Fig. 7.18 Room temperature Mössbauer spectrum of allanite. [W. A. Dollase, *Amer. Mineral.*, **56**, 447 (1971).]

the M_3 position. The single doublet observed in the Mössbauer spectrum (Table 7.1) was assigned to Fe^{3+} in M_3. Optical spectra of Mn piemontites suggested that there was appreciable $Mn^{3+} + Fe^{3+}$ in the M_2 position, but X-ray evidence shows that $Mn^{3+} + Fe^{3+}$ enters M_1 rather than M_2.

In allanite, the X-ray and chemical data indicate that Fe^{2+} and Fe^{3+} can only be in M_1 and M_3 giving a total of eight possible peaks in the Mössbauer spectrum: two Fe^{2+} and two Fe^{3+} doublets. The spectrum (Fig. 7.18) shows two well resolved doublets, which from their centre shifts (Table 7.10) can be assigned: A and A' to Fe^{2+}, and C and C' to Fe^{3+}. Two other weak doublets were fitted to this spectrum, although no statistical reasons were given for doing so. From the centre shifts, B and B' can be assigned to Fe^{2+}, and D and D' to Fe^{3+}. Since the parameters for C and C' are in good agreement with previous parameters observed for epidotes (C.S. = 0.61 mm s^{-1}; Q.S. = 2.01 mm s^{-1}), this doublet can be assigned to Fe^{3+} in M_3. It follows from the chemical and crystallographic data that D and D' are due to Fe^{3+} in the less distorted M_1 position. The smaller Fe^{2+} quadrupole splitting for peaks A and A' suggests assignment of this doublet to the distorted M_3 position, since the Fe^{2+} quadrupole splitting decreases as the distortion

Table 7.10 Mössbauer parameters (mm s^{-1}) and assignment of peaks for allanite

Line pair	C.S.	Q.S.
A	1·33 (3)	1·66 (4)
B	1·50 (15)	1·93 (21)
C	0·63 (1)	1·97 (3)
D	0·55 (7)	1·33 (12)

W. A. Dollase, *Amer. Mineral.*, **56**, 447 (1971).

from octahedral symmetry increases (chapter 6 and section 5 of this chapter). This leaves B and B′ to Fe^{2+} in M_1. The site populations derived on the basis of this assignment were reasonably consistent with chemical analysis and X-ray diffraction data.

7.4 Line widths and areas

Using a ^{57}Co in Pd or Cu source, most minerals discussed in this chapter give line widths of less than 0·35 mm s^{-1}, if the line is due to a single type of iron atom. The most common kind of line broadening is caused by an overlap of two or more non-coincident Lorentzians as noted previously for such spectra as deerite and the cummingtonite–grunerites. However, as noted for the Kakanui augite, line widths for a single type of iron atom were very broad (>0·45 mm s^{-1}). There appear to be two mechanisms which could give rise to such line broadening. The first, and most probable, might be termed the randomization mechanism. One Fe^{2+} atom (A) in M_1 might have an Fe^{2+} next to it in M_2, while another Fe^{2+} (B) in M_1 might have a Ca^{2+} adjoining it in M_2. A and B will thus see a slightly different electric field gradient, and give slightly different quadrupole splittings. As discussed earlier, this type of mechanism may be the correct explanation for the anomalous augite spectra (section 7.3). A second mechanism which might give line broadening is a random strain, such that the bond lengths varied significantly from site to site, perhaps due to small variations in chemical composition. Relaxation effects can also give line broadening, but these effects are not important for mineral spectra taken above 80 K. Despite the above possible broadening mechanisms, if a line is asymmetric, it seems almost certain that resolution into two or more component peaks is possible; if broad, (>0·35 mm s^{-1}), it is likely that the sample is not homogeneous, or that one of the above mechanisms is operating.

The line widths of component doublet peaks can normally be assumed equal, but care must be taken when constraining all line widths to be equal (as in the omphacites). If a fit will not converge without such a constraint,

Mössbauer spectroscopy

then it is worth using. However, the overall confidence in the peak parameters decreases.

The areas of component peaks of a doublet for a random sample are often very closely equal; but, as pointed out for many spectra in this chapter, the low velocity peak is often significantly more intense than the component high velocity peak. It is often necessary in very complex spectra to constrain the peak areas to be equal, and so it is desirable usually to have the areas equal. There are two possible mechanisms which cause the area ratio to deviate from 1:1, and these have been mentioned in section 2.3: an oriented sample, and the Goldanskii effect. Methods for obtaining random samples have been mentioned in section 3.2, but the Goldanskii effect cannot be eliminated. It is a property of the mineral, and will generally increase as the site distortion from octahedral symmetry increases.

7.5 Correlation of quadrupole splitting with structural variations

As seen in Fig. 7.1b, there is a very large variation in iron quadrupole splittings. Ferric quadrupole splittings vary from the aegirine–jadeite value of 0.33 mm s^{-1} (Table 7.4) to 2·0 mm s^{-1} for epidotes (Table 7.10); and ferrous quadrupole splittings range from a low of 0·51 mm s^{-1} for gillespite to a high of 3·56 mm s^{-1} for garnets. In addition, as indicated in Tables 7.4 and 7.7, there is considerable variation in quadrupole splitting within one solid solution series. For example, the M_4 quadrupole splitting in the cummingtonite–grunerite series varies from 1·64 mm s^{-1} for a cummingtonite, to 1·50 mm s^{-1} for a grunerite. These variations in quadrupole splitting can be related in a very qualitative manner at the present time to structural effects.

As discussed in section 2.2, the field gradient may be expressed as:

$$q = (1 - \gamma_{\text{infinity}})q_{\text{lattice}} + (1 - R)q_{\text{valence}} \qquad (2.12)$$

For Fe^{3+} minerals, we can take $q_{\text{valence}} = 0$, and the quadrupole splitting is due to q_{lattice} (Table 2.2). The Fe^{3+} quadrupole splitting would then be expected to *increase* as the distortion of the atoms from octahedral symmetry about Fe^{3+} *increases*. The small quadrupole splitting in aegirine–jadeite might then be expected from the nearly octahedral symmetry of the oxygen environment (Table 7.3). Similarly, the very large quadrupole splitting in epidote can be attributed to the very distorted M_3 site in epidotes.

In contrast to the above situation, Fe^{2+} gives rise to a large q_{valence} term (section 2.2) when there is a very small distortion from octahedral symmetry (Fig. 6.9). A very large quadrupole splitting of about 3·7 mm s^{-1} results. As the distortion of the ligand environment from octahedral increases, q_{lattice} increases, but the quadrupole splitting *decreases* from the q_{valence} value (section 2.2). Thus, in the relatively symmetric M_1 site in amphiboles (Table

7.6), a large splitting of $\sim 2\cdot 8$ mm s^{-1} is observed (Table 7.7), while in the very distorted square planar environment (one of the limiting cases of a tetragonal distortion) in gillespite, a very small quadrupole splitting of 0·51 mm s^{-1} is observed.

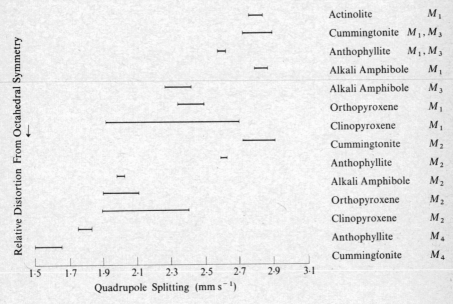

Fig. 7.19 Quadrupole splitting plotted versus distortion of the coordinating oxygen polyhedra from octahedral symmetry. Adapted from G. M. Bancroft, A. G. Maddock and R. G. Burns, *Geochim. Cosmochim. Acta*, **31**, 2219 (1967).

In Fig. 7.19, we plot the qualitative relative distortion of the Fe^{2+} − O polyhedra versus quadrupole splitting for a large number of silicates. Very broadly, the quadrupole splitting again decreases as the distortion increases. This type of correlation is useful when trying to assign Fe^{2+} to different positions in silicate structures. For example, Fe^{2+} in the very distorted pyroxene M_2 and amphibole M_4 sites would be expected to give smaller quadrupole splittings than Fe^{2+} in the pyroxene M_1 and amphibole M_{123} sites respectively, and this is observed. Similarly, crystal structures of orthopyroxenes, cummingtonite and grunerite indicate that the M_2 and M_4 oxygen polyhedra become more distorted with increasing Fe^{2+} content. (Tables 7.3 and 7.6). Figure 7.20 shows that the quadrupole splitting decreases with increasing Fe^{2+} content as we might expect from the above distortion criteria.

However, the above treatment is not an accurate guide to many quadrupole splittings. For example, the pyroxene M_1 sites (Table 7.3) show comparatively

Mössbauer spectroscopy

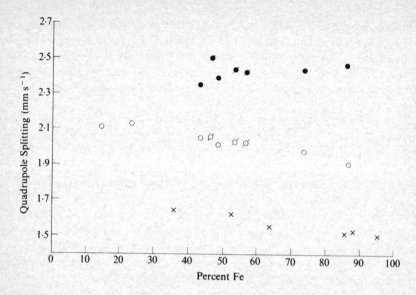

Fig. 7.20 Variation of the room temperature quadrupole splitting within the cummingtonite and orthopyroxene phases. ● Metamorphic orthopyroxene M_1 sites; ● volcanic orthopyroxene M_1 sites; ○ metamorphic orthopyroxene M_2 sites; ○ volcanic orthopyroxene M_2 sites; × cummingtonite-grunerite M_4 site. [G. M. Bancroft, A. G. Maddock, and R. G. Burns, *Geochim. Cosmochim. Acta*, **31**, 2219 (1967); S. Ghose and S. S. Hafner, *Zeit. Krist.*, **125**, 157 (1967).]

little variation in distortion, yet the M_1 quadrupole splitting varies by almost 1 mm s^{-1} (Table 7.4). Similarly, the amphibole M_1 and M_3 sites (Table 7.6) show little variation in distortion of the oxygen polyhedra, and yet in actinolite and other alkali amphiboles, there is a difference in quadrupole splitting of about 0.4 mm s^{-1}.

A number of other effects would be expected to contribute to $q_{lattice}$, besides the distortion of the oxygen polyhedron. First, the effective charges on the oxygen atoms will vary markedly, because some oxygens are linked to one silicon, some to two, and some are common to two or more polyhedra. In addition in amphiboles, some oxygens are found in OH groups. Second, we have neglected contributions to $q_{lattice}$ from other than nearest neighbour oxygens. The symmetry of and charge on other neighbouring cations, silicons and oxygens must markedly affect the quadrupole splittings. For example, substitution of Ca^{2+} for Fe^{2+} in the orthopyroxene M_2 and cummingtonite M_4 positions influences the quadrupole splitting markedly in the neighbouring positions. Thus, in hedenbergite (where Ca^{2+} replaces Fe^{2+} in the M_2 position), the M_1 position gives rise to a much smaller quadrupole splitting than in the orthopyroxene M_1 position (Table 7.4);

similarly, in actinolite (where Ca^{2+} replaces Fe^{2+} in the M_4 position), the M_2 position gives rise to a quadrupole splitting about 0·9 mm s^{-1} smaller than in the cummingtonite M_2 position (Table 7.7). Unfortunately, the above effects are difficult to estimate—even qualitatively—and the distortion criteria appears to be generally useful when applied with some caution.

The temperature dependence of the quadrupole splittings is also related to the magnitude of $q_{lattice}$ (section 2.2) and the resulting splitting of the low energy $t_{2g}\,d$ orbitals. If the distortion from octahedral symmetry is small, then Δ_3 is small (eq. 2.17 and chapter 6.5), and $q_{valence}$ increases markedly as the temperature decreases. Thus, the pyroxene and amphibole M_1 positions give rise to temperature dependent quadrupole splittings. An increase in the quadrupole splitting of about 0·5 mm s^{-1} from 295 K to 80 K is common. If $q_{lattice}$ and Δ_3 are large (as in the M_2 pyroxene and M_4 amphibole positions), then a temperature independent quadrupole splitting is observed. As $q_{valence}$ remains constant, any temperature variation of the quadrupole splitting (as for Fe^{3+} minerals) is due to variations in bond lengths.

The above variation in the temperature dependence of the quadrupole splitting indicates that the resolution of complex spectra will vary markedly with temperature. For example, the orthopyroxene M_1 and M_2 doublets are better resolved at 80 K than 295 K. For many spectra the temperature of best resolution has not yet been obtained.

7.6 Uses of characteristic Fe^{2+} and Fe^{3+} peaks in solid state processes

In the preceding sections of this chapter, we have been concerned with spectral features such as area and line widths, and the assignment of peaks to Fe^{2+} in various cation positions. In the next chapter, quantitative applications of Mössbauer areas, and possible applications in geothermometry will be discussed.

Besides the purely analytical function of Fe^{3+}/Fe^{2+} ratios, we indicated earlier that these ratios have been employed as indications of the temperature of formation of minerals, and that there is a possibility that the ratio could be used as an indication of pressure of formation when combined with such high pressure laboratory experiments described in section 4.4. In this section, we will discuss several other fundamental solid state processes in minerals which rely on peak assignments previously discussed: e.g. colour and pleochroism, and oxidation and weathering processes in minerals. The latter applications will probably be among the most important long term Mössbauer applications in geochemistry, but very few such studies have yet been reported. The non-destructive nature of the experiment makes it ideally suited for studying the above, and other, solid-state processes.

Mössbauer spectroscopy

Charge transfer spectra, colour and pleochroism. Very intense absorption bands in the visible or ultraviolet portion of the electromagnetic spectrum are often due to what are known as charge transfer bands. These charge transfer bands are normally very important in determining the colour, and pleochroism of minerals containing transition metals. These bands are due to electron migration *between* atoms or ions, as opposed to the electronic transitions described in the next chapter which take place *within* the d levels of one transition metal atom. The charge transfer transitions may take place from metal to ligand, ligand to metal, or metal to metal, and they are characterized by the very intense bands, and resulting intense colour of the mineral (ref. 7.8).

Charge transfer transitions are favoured when neighbouring ions in a structure are in different oxidation states, e.g. Fe^{2+} and Fe^{3+}; and/or when there is a local charge misbalance due, for example, to substitution of a divalent ion for a trivalent ion or vice versa. In order to explain the intensity and polarization dependence of a charge transfer band, it is essential to know the approximate amount of each type of ion in each structural position. The

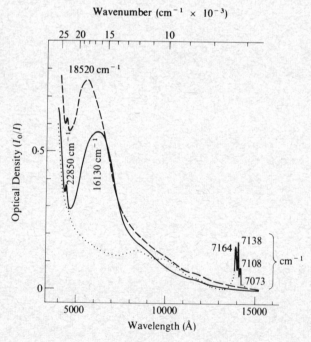

Fig. 7.21 Polarized absorption spectra of glaucophane. .... α spectrum; ---- β spectrum; ——— γ spectrum (optic orientation: α: a = 10°; β = b; γ: c = 4°). R. G. Burns, *Mineralogical Applications of Crystal Field Theory*, Cambridge University Press, 1970, Figure 4.14.

assignments discussed earlier for alkali amphiboles will illustrate the relationship between structure, colour, and pleochroism.

Alkali amphiboles of the glaucophane–riebeckite series [$Na_2Mg_3Al_2Si_8O_{22}(OH)_2$ to $Na_2Fe_3^{2+}Fe_2^{3+}Si_8O_{22}(OH)_2$] are characterized by their distinctive colourless–violet–blue pleochroism in transmitted polarized light with $E \sim //a$, $E//b$ and $E \sim //c$ respectively. A polarized absorption spectrum of a glaucophane (approximate composition $Na_2(Mg_2Fe_{1.0}^{2+}Fe_{0.5}^{3+}Al_{1.5})Si_8O_{22}(OH)_2$) is shown in Fig. 7.21. Light polarized approximately along a shows little absorption in the visible region (colourless), while light polarized along b and c gives intense absorptions which lead to the violet–blue pleochroism. The intensity of these absorption bands increases with rising Fe^{2+} and Fe^{3+} in these minerals.

The absorption maxima in Fig. 7.21 can be assigned to charge transfer bands arising between Fe^{2+} and Fe^{3+} ions in the structure. The cation distribution (discussed earlier in section 7.3), partially deduced from Mössbauer spectra, is summarized in Fig. 7.22. Sodium ions fill the M_4 positions, Fe^{2+} and Mg^{2+} are concentrated in M_1 and M_3, while Fe^{3+} and Al^{3+} prefer the M_2 position. Electron transfer takes place most favourably when light is polarized $E//b$, since the Fe^{2+} and Fe^{3+} ions lie in adjacent sites

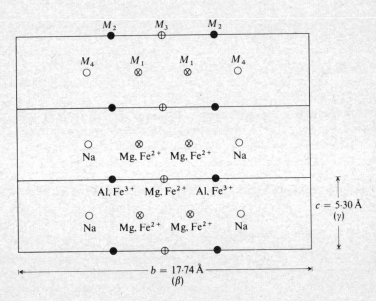

Fig. 7.22 Portion of the glaucophane structure projected onto (100). The cation positions are indicated in the top unit cell. The middle unit cell shows the cation distribution in the glaucophane structure. [G. M. Bancroft and R. G. Burns, *Min. Soc. Amer. Spec. Pap.*, **2**, 137 (1969).]

along the b axis. A slightly smaller interaction takes place with $E \sim //c$ since the Fe^{3+}-Fe^{2+} pairs in M_2 and M_3 respectively are inclined at an angle to the c axis. Thus the intensity of the $E//b$ spectrum is greater than for the $E//c$ spectrum. Little or no absorption takes place for $E//a$, since the cations are shielded from each other by the silicate framework (Fig. 7.13). As the number of Fe^{3+}–Fe^{2+} pairs increases, the intensity of the bands and the intensity of the colour of the mineral increases.

Similar effects in other minerals which contain Fe^{2+} and Fe^{3+} are also known. Vivianite, $Fe_3(PO_4)_2 \cdot 8H_2O$ presents a dramatic example of the development of a deep colour on attainment of mixed valence. The pure mineral (containing only Fe^{2+}) is colourless until exposed to air whereupon a small amount of Fe^{3+} is produced (shown by characteristic Fe^{3+} Mössbauer peaks) and the mineral becomes an intense blue colour. This blue colour can again be attributed to a charge transfer band in the red end of the spectrum, with the blue light being transmitted.

Oxidation and weathering. Oxidation and weathering processes in minerals are of great geological interest as well as of considerable industrial importance. The mechanisms of these processes are largely unknown. Preliminary work on the oxidation of crocidolite (Table 7.5) suggested that the oxidation to oxycrocidolite $Na_2Fe_4^{3+}Fe_{0.6}^{2+}Mg_{0.4}Si_8O_{22}(OH)_2$ proceeded by a mechanism of electron and proton transfer along the silicate chain rather than across the silicate chains. The charge transfer processes described above for alkali amphiboles have the same directional characteristics. The Mössbauer study on the oxidation of crocidolite did not add significantly to the knowledge about the above mechanism.

However, the Mössbauer study on the oxidation of amosite (fibrous grunerite) lead to more interesting results. The ultimate decomposition products of amosite had been shown to be pyroxene and water, together with some cristobalite. The Mössbauer spectrum of an amosite of approximate composition $(Fe_{5.5}Mg_{1.5})Si_8O_{22}(OH)_2$ (Fig. 7.23a) is very similar to that of the grunerite shown earlier (Fig. 7.14c). The peaks can be assigned as previously: inner two peaks to Fe^{2+} in M_4, and outer two to Fe^{2+} in M_{123}. On heating in air at 500°C, peaks characteristic of Fe^{3+} appear (C and D, Fig. 7.23b), but it is noticeable that peak E (due to Fe^{2+} in M_4) is becoming relatively more intense than peak F (due to Fe^{2+} in M_{123}). Oxyamosite, having an approximate composition $Fe_2^{3+}Fe_{3.5}^{2+}Mg_{1.5}Si_8O_{22}$, is formed. Slightly more Fe^{3+} is formed at 650°C, and again the M_4 peak becomes relatively less intense. The quadrupole splittings for the M_4 and M_{123} peaks remain similar, but the M_{123} quadrupole splittings increase slightly during the heating process.

The above spectra show that Fe^{2+} in M_{123} is oxidized rather than Fe^{2+} in M_4. Also, the similar Mössbauer parameters for amosite and oxyamosite suggest that the basic structure, and the coordination about the metal cation

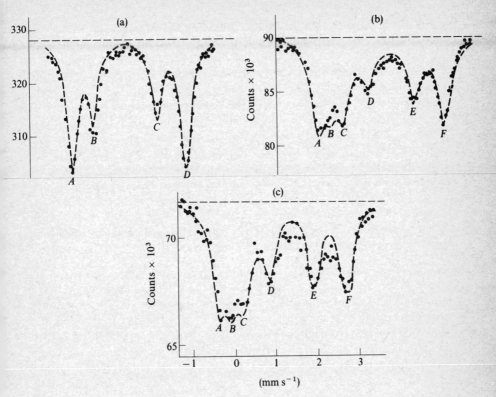

Fig. 7.23 Mossbauer spectra of: (a) amosite; (b) amosite heated to 500°C (oxyamosite I); (c) amosite heated to 650°C (oxyamosite II). [H. J. Whitfield and A. G. Freeman, *J. Inorg. Nucl. Chem.*, **29**, 903 (1967).]

sites remains the same. These results are consistent with the previous suggestion for crocidolite that there is no electron transfer between cation chains, but only along cation chains.

At 900°C, a doublet is observed with a quadrupole splitting of 2·15 mm s^{-1} indicating that a Fe^{2+} pyroxene is formed (Table 7.4). In addition, six broad peaks appeared due to a mixture of hematite and the spinel (Mg, Fe^{2+})-$Fe_2^{3+}O_4$.

The weathering of micas is an important geochemical process. Iron in fresh biotites is predominantly in the ferrous form (Fig. 7.6), but natural weathering processes in soils or fractured rocks oxidize the ferrous iron to Fe^{3+}. This oxidation is accompanied by other changes in the biotite which can be formulated by first, loss of interlayer cations:

i.e. $$KFe_3^{2+}Si_3AlO_{10}(OH)_2 + \tfrac{1}{4}O_2 + \tfrac{1}{2}H_2O$$
$$\rightarrow KOH + Fe_2^{2+}Fe^{3+}Si_3AlO_{10}(OH)_2$$

or second, loss of protons from OH:

i.e.

$$KFe_3^{2+}Si_3AlO_{10}(OH)_2 + \tfrac{1}{2}O_2 \rightarrow KFe^{2+}Fe_2^{3+}Si_3AlO_{10}(O_2) + H_2O$$

or third, loss of octahedral iron:

i.e.

$$KFe_3^{2+}Si_3AlO_{10}(OH)_2 + \tfrac{3}{4}O_2 + \tfrac{1}{2}H_2O \rightarrow KFe_2^{3+}Si_3AlO_{10}(OH)_2 + FeOOH$$

Mössbauer spectroscopy would appear to be an ideal technique for studying such weathering processes, but results to date have not fulfilled the potential of Mössbauer in such studies. The spectra of weathered biotites (ref. 7.9) clearly show that Fe^{2+} is oxidized to Fe^{3+}, but the spectra were not good enough to show whether FeOOH (which gives a characteristic six-line pattern) was formed, as indicated by electron microscopy and X-ray work. Also the two ferrous doublets were not resolved (as in Fig. 7.17a), so that information on the relative ratio of oxidation of M_1 and M_2 Fe^{2+} could be obtained. One weathering study has reported the formation of low spin Fe^{III} as well as Fe^{3+}, but this assignment is somewhat questionable because of the poor resolution of these peaks.

Interestingly, resaturation of the potassium-depleted products with K^+ left the Fe^{3+}/Fe^{2+} ratio (determined by Mössbauer) essentially unchanged, and the amount of oxidation appears to be independent of the total iron content, or type of mica. Iron oxidation apparently occurs independently of other interlayer alteration, and requires just that O_2 reaches the iron sites. Once the interlayers have been expanded and the iron oxidized, there appears to be no further change in the Fe^{3+}/Fe^{2+} ratio upon reintroducing K^+.

A Mössbauer study of natural ilmenites ($FeTiO_3$) (ref. 7.10) also points out the possibilities of using Mössbauer spectra in weathering studies. The natural ilmenites gave spectra containing not only the doublet characteristic of $FeTiO_3$(C.S. = 1·35 mm s^{-1}; Q.S. = 0·70 mm s^{-1}), but a ferric species (C.S. = 0·62 mm s^{-1}; Q.S. = 0·67 mm s^{-1}) which could not be characterized, but was attributed to a weathering product. The degree of conversion, and thus the degree of weathering, can readily be determined from the characteristic Mössbauer spectra.

Problems

1. Small amounts of iron are often present in natural feldspars, but it is very difficult to assess the oxidation state of Fe, and its position in the feldspar structure. In lunar plagioclases ($NaAlSi_3O_8$–$CaAl_2Si_2O_8$), ^{57}Fe

Mössbauer spectra have been obtained [S. S. Hafner, D. Virgo and D. Warburton, *Earth Planetary Science Lett.* **12**, 159 (1971)]. These spectra consist of two doublets with centre shifts of 1·39 mm s^{-1} and 1·16 mm s^{-1} respectively, and quadrupole splittings of 2·01 mm s^{-1} and 1·54 mm s^{-1} respectively at 295 K. What type of Fe could these two doublets be assigned to, and what positions in the structure do the iron atoms likely occupy?

2. The Mössbauer spectra of synthetic ferridiopsides ($\sim$(Ca, Mg, Fe^{3+})$_2$-Si$_2$O$_6$) consist of two doublets having centre shifts of 0·40 mm s^{-1} and 0·68 mm s^{-1} at 295 K. Discuss possible oxidation state and structural assignments of these two doublets.

3. In the spectrum below, how would you pair off the four peaks? How could you confirm your assignment? What are likely causes of the different peak heights within one doublet? (E. Kostiner, *Amer. Min.*, **57**, 1109, (1972)).

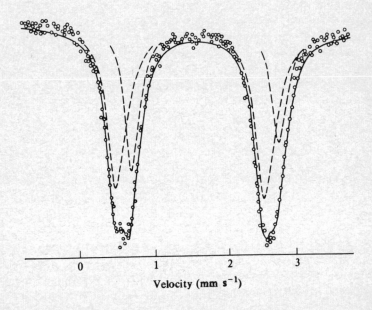

4. Herzenbergite [formally Sn(SnS$_2$)] gives a two line ^{119}Sn Mössbauer spectrum: a strong line with position 0·00 mm s^{-1} and a weaker line with position 3·14 mm s^{-1}. The area ratio of these peaks is strongly temperature dependent. Give possible interpretations of the spectrum. Use the centre shift information for Sn near the beginning of chapter 4.

References

7.1 U. Gakiel and M. Malamud, *Amer. Mineral.*, **54**, 299 (1969).
7.2 G. M. Bancroft, R. G. Burns and A. G. Maddock, *Acta. Cryst.* **22**, 934 (1967).
7.3 R. G. Strens, *Chem. Comm.*, 777 (1966).
7.4 D. J. Elias and J. W. Linnett, *Trans. Farad. Soc.*, **65**, 2673 (1969).
7.5 F. F. Brown and A. M. Pritchard, *Earth Planet Sci. Lett.*, **5**, 259 (1968).
7.6 J. R. Clark, D. E. Appleman and J. J. Papike, *Mineral Soc. Amer. Spec. Pap.* **2**, 31 (1969) and references.
7.7 S. S. Hafner and H. G. Huckenholz, *Nature* **233**, 9 (1971).
7.8 R. G. Burns, *Mineralogical Applications of Crystal Field Theory*, Cambridge University Press (1970).
7.9 L. H. Bowen, S. B. Weed and J. G. Stevens, *Amer. Mineral.* **54**, 72 (1969).
7.10 T. C. Gibb, N. N. Greenwood and W. Twist, *J. Inorg. Nucl. Chem.*, **31**, 947, 947.

Bibliography

A. G. Maddock, in I.A.E.A. reviews (1972), p. 329.
W. A. Deer, R. A. Howie and J. Zussman, *Rock-Forming Minerals*, Longmans (1963).

8. Quantitative site populations in silicate minerals

As we have seen in the previous chapter, silicate minerals generally have two or more cation sites into which Ca^{2+}, Mg^{2+}, Fe^{2+}, Al^{3+}, and Na^+ can enter. Often the cation sites are of similar size and energy, and one cation such as Fe^{2+} occupies more than one position, although it will usually prefer to enter one of the cation positions—a process called ordering. The extent of ordering might be expected to be dependent on the temperature and pressure of formation e.g., as the temperature of formation increases, the cation distribution should become more random. Thus knowledge of the quantitative cation distributions in silicates may be a useful indicator of the thermal history of these minerals.

Up until the last four or five years, very little data on cation ordering in silicates had been obtained. Chemical analyses together with optical and density measurements may suggest ordering, but site populations cannot be estimated. Using X-ray diffraction, ordering had been detected in a number of silicates, but until about seven years ago, site populations had only been estimated in one alkali amphibole, one cummingtonite, one hypersthene, and one pigeonite. However, with the recent advances in X-ray technology, many structures and site populations have been determined by X-ray methods in the last few years (ref. 8.1). The X-ray method is still relatively time consuming, and it is impossible at the present time to distinguish between ions having very similar scattering factors such as Fe^{2+}, Fe^{3+}, and Mn^{2+}.

In 1966, reports of two spectroscopic methods being applied to quantitative estimation of cation distributions were first made. The first—infrared spectroscopy (ref. 8.2)—may indicate ordering, but it does not yet appear to be a precise or accurate method, and it is limited to silicates containing hydroxyl groups. The Mössbauer method (ref. 8.3) described in the following pages appears to give rapid and accurate estimates of Fe^{2+} site population ratios and Fe^{2+}/Fe^{3+} ratios in silicate minerals. A large amount of recent data enables us to evaluate the method critically.

Mössbauer spectroscopy

8.1 The Mössbauer method for determining site population ratios

If a mineral contains two different types of iron atoms A and B which give rise to *non-overlapping* Mössbauer spectra, the area ratio for a *thin* absorber can be expressed as:

$$A_A/A_B = C N_A/N_B \qquad (8.1)$$

where:

$$C = \frac{\Gamma_A G(X_A) f_A}{\Gamma_B G(X_B) f_B} \qquad (8.2)$$

and Γ = width at half height
$G(X)$ = saturation corrections (eq. 2.26)
f = recoil free fractions
N_A = number of iron atoms per formula unit in site A.
$N = kn$ (eq. 2.26) where k is a constant.

Since $N = kn$, $N_A/N_B = n_A/n_B$. The usual error associated with measuring areas—background—is minimized in the area ratio, since the background corrections for the two iron atoms are similar. The areas can be determined directly from the spectra, and if C can be determined or estimated, the *ratio* of iron atoms in the two different sites can be determined. The constant C can be determined from a mineral in which N_A/N_B is already known. For example, in the orthopyroxene series, C could be determined from the iron end member ferrosilite, $Fe_2Si_2O_6$, in which N_2/N_1 is, of course, one. When dealing with a series of minerals such as orthopyroxenes, it is then assumed that C is a constant for all $Fe^{2+}:Mg^{2+}$ ratios. Knowing the areas A_2 and A_1 for any orthopyroxene, it is then possible to determine N_2/N_1 using eq. (8.1). If N_1 plus N_2 (the total Fe^{2+} content in this case) is known from chemical analyses, the site populations N_1 and N_2 can then be calculated.

Unfortunately, for most silicates it is difficult to obtain an accurate estimate of C. The f factors can be determined by several methods (see eq. (2.25) for one), but they are rather laborious and the rather large errors in the determination make it questionable whether it is worth doing. For most silicates, we assume that $C = 1$ and the area ratio is then equal to the ratio of the number of iron atoms. This implies that:

$$\Gamma_A = \Gamma_B; \qquad f_A = f_B; \qquad G(X_A) = G(X_B)$$

The widths of most mineral peaks have been found to be very closely equal if the same source is used (chapter 7). The f factors should be very closely equal—especially if the spectra are taken at the lowest possible temperature, and the saturation corrections $G(X)$ approach one if thin absorbers are used. Thin here implies that $X(= nf_a\sigma_o) \leq 0.2$, giving $G(X) \geq 0.96$, where n is the number of ^{57}Fe atoms per unit area of the absorber, and σ_o is the maximum cross section at resonance = $2 \cdot 35 \times 10^{-18}$ cm^2 (eq. 2.27). $G(X)$ decreases from 1 as X increases. For example, $G(X) \simeq 0 \cdot 80$ when $nf_a\sigma_o = 1$. To calculate n for $X = 0 \cdot 2$, we take $f_a = 0 \cdot 6$, and n becomes $1 \cdot 4 \times 10^{-2}$ mg cm^{-2} of ^{57}Fe. Remembering that ^{57}Fe occurs in only about two per cent natural abundance, the above amount of ^{57}Fe corresponds to $0 \cdot 7$ mg cm^{-2} of natural iron, or about 7 mg cm^{-2} of a mineral containing 10 wgt % iron. With iron rich pyroxenes or amphiboles for which the weight percentage of Fe may be over 30%, then a smaller amount of mineral should be used. However, it should be remembered that C includes the *ratio* of the G's, and as long as the amount of Fe in sites A and B are similar, then even if the G's deviate markedly from one, the ratio still will be close to one. As a general rule, it is best to use less than 5 mg cm^{-2} of mineral in an absorber.

Since the Mössbauer experiment only 'looks at' ^{57}Fe, and it occurs in only about two per cent natural abundance, it should be kept in mind that we tacitly assume that no isotope fractionation takes place between sites. There appears to be no evidence for isotope fractionation.

There are also other important possible errors associated with estimating areas. First, eq. (8.1) only holds for thin absorbers with lines that do not overlap. Inevitably, mineral absorption lines *strongly* overlap (Figs. 3.7, 7.8, 7.14, 7.15), and a detailed area treatment of such cases—even when the lines can be 'resolved' using a computer—has not yet appeared. Later results will indicate that at least for fairly well resolved spectra, such as cummingtonite-grunerite (Fig. 7.14) and most orthopyroxene spectra at liquid N$_2$ temperatures, very small errors will be encountered because of line overlap.

Second, as discussed in chapter 2, a Lorentzian line shape is assumed, so that any deviation of lines from Lorentzian shapes will of course affect the estimated area. The line shapes appear to be very close to Lorentzian, but deviation from this shape could be caused by a thick absorber ($X \gg 1$), a strong overlap of a number of unresolvable Lorentzians, instrumental effects, or more subtle relaxation effects. The only one of general importance is the close overlap of two or more Lorentzians. For example in the cummingtonite–grunerite spectra (Fig. 7.14), the outer two lines are due to a superposition of three unresolvable Lorentzians due to Fe^{2+} in the M_1, M_2 and M_3 positions. These outer lines will not then be *exactly* Lorentzian.

The third important problem concerning estimation of areas is a practical one. To 'resolve' two Lorentzians (using a computer) and obtain accurate areas, the two peaks must be separated by at least the half width at half height—normally about $0 \cdot 15$ mm s^{-1} for mineral spectra. If the peaks are $\sim 0 \cdot 20$ mm s^{-1} apart (as in room temperature spectra of orthopyroxenes containing less than 50% Fe^{2+} per formula unit) the standard deviations in the areas are very large ($>15\%$) because the computer is just able to resolve the peaks. If the separation of the peaks is ≥ 0.30 mm s^{-1}, the statistical errors become more reasonable for an average spectrum ($< 10\%$) and will decrease markedly as the quality of the spectrum increases. It seems important then to quote standard deviations in the areas and/or some measure of the reproducibility of the areas.

It is often necessary to impose constraints on the fitting procedure especially in the initial stages—in order that convergence occurs. Either half widths or intensities can be constrained to be equal, but when evaluating the accuracy of the areas derived from such fits, it should be remembered that the half widths—or areas—may not be *exactly* equal.

Although the area ratios for most silicates now can be derived with considerable precision ($\pm 1\%$), the accuracy of the results is still a matter of some conjecture, because of the above assumptions, and some chemical problems mentioned in the next section. In evaluating the accuracy of the method, it seems important to compare results obtained by Mössbauer and other methods such as X-ray diffraction and chemical analysis. These comparisons indicate that the accuracy of the site populations derived is better than $\pm 5\%$.

8.2 Site populations and comparisons with X-ray and chemical analyses values

In deriving Fe^{2+} site populations in silicates, it is necessary to know not only the site population ratio N_A/N_B derived from eq. (8.1) and Mössbauer areas, but the total ferrous concentration. Similarly, to obtain Fe^{2+} and Fe^{3+} concentrations, it is necessary to know the total iron content. These values are normally obtained by wet chemical analyses, or by the electron microprobe. Any errors in analyses will be reflected in the site populations. To ensure that the sample is homogeneous, examination of several crystals by the electron microprobe is desirable. In some minerals, it is possible to have exsolution or fine zoning (e.g. lunar pyroxenes) which, if extensive, will make any derived site populations much less meaningful.

In evaluating the significance of the ferrous site populations, it is important to know (from chemical analyses) what amounts of other cations are present. In a binary solid solution such as orthopyroxenes [(Fe, Mg)$_2$Si$_2$O$_6$] and cummingtonite–grunerite [(Fe, Mg)$_7$Si$_8$O$_{22}$(OH)$_2$], determination of the ferrous site populations immediately enables the calculation of the magnesium site populations by difference. However, invariably there are small amounts of other cations (e.g. Ca^{2+} and Mn^{2+}), which may order in one of the sites and markedly alter the *apparent* extent of ferrous ordering. For example, consider two hypothetical orthopyroxenes X and Y both having ferrous site populations of 0.8 Fe^{2+} per formula unit in M_2 and 0.20 Fe^{2+} per formula unit in M_1. Sample X has, in addition to 0.90 Mg^{2+}, 0.10 Ca^{2+} which is assigned to the M_2 position; while sample Y has 1.00 Mg^{2+} and no Ca^{2+}. Thus, Fe^{2+} fills 89% of the available M_2 sites in sample X, but only 80% of the available

M_2 sites in Y, although the ferrous site populations in the two samples are identical. Ca^{2+} can usually be safely assigned to the largest of the available sites, but other ions such as Mn^{2+} will distribute between the available sites. However, in orthopyroxenes, any Ca^{2+} may be present as exsolved clinopyroxene. Because of these problems, some workers have decided that the Fe^{2+}, Mg^{2+} site occupancy is best represented on the basis of the simple binary Fe, Mg solid solution until additional data on the exsolution effects and the site populations of all elements are available. These problems with chemical analyses and exsolution are also important to the X-ray crystallographer, but Mössbauer spectra require a large number of crystals in an absorber, all of which cannot be analyzed and examined.

Despite the above list of possible difficulties in obtaining site populations, a comparison of site populations (Table 8.1) determined by both X-ray and Mössbauer methods demonstrate very good agreement, and lend very strong support not only to the Mössbauer assumptions, but also to the X-ray data.

The consistently good agreement in Table 8.1 is perhaps surprising because of the different methods used to calculate these values for very different types of spectra. The value of C was taken to be one for specimens 1, 2, and 3, while the calculated value of $C = 0.90$ (vide infra) for grunerite was used for samples 4, 5, and 6. With samples of 5–10 mg cm^{-2}, $G(X_A)/G(X_B)$ would be expected to be significantly different from one for all specimens but 2 and 3, because A_A/A_B (and N_A/N_B) is very different from one for specimens 1, 4, 5, and 6. The peaks in specimens 2, 4, and 6 are comparatively well resolved (e.g., see Fig. 7.14), while for specimens 1, 3, and 5 (e.g., see Fig. 7.15), no more than distinct shoulders are apparent in the ferrous envelope.

These data certainly indicate that the strong overlap of lines and saturation corrections lead to small errors. The differences in C values, however, need closer examination.

The Mössbauer and chemical analyses values for ferric to ferrous ratios (Table 8.2) are not in quite as good agreement. Because of the difficulty in dissolving some minerals, with resulting oxidation, the chemical analyses values often do not reflect the chemical content. Several large discrepancies between Mössbauer results and chemical analyses have been resolved with a new chemical analysis which has been in substantial agreement with the Mössbauer figures. Because of the non-destructive nature of the Mössbauer technique, it would seem to be an excellent method for the routine determination of accurate Fe^{3+} to Fe^{2+} ratios.

In conclusion, present results indicate that the Mössbauer method provides a rapid and accurate method for quantitative determination of ferric to ferrous ratios and ferrous site population ratios. When used on its own for binary solid solutions, or in conjunction with other techniques for minerals containing several cations, it will be possible in the next few years to derive a

Mössbauer spectroscopy

Table 8.1 Comparison of X-ray and Mössbauer site populations in silicate minerals

Mineral	Total Fe^{2+} per formula unit	Fe^{2+} site populations (per formula unit)	
		Mössbauer (a)	X-ray (b)
1. Orthopyroxene	1·06	$M_2 = 0·87$	$M_2 = 0·90$
		$M_1 = 0·19$	$M_1 = 0·15$
2. Orthopyroxene	1·70	$M_2 = 0·95$	$M_2 = 0·96$
		$M_1 = 0·75$	$M_1 = 0·74$
3. Glaucophane	0·61	$M_3 = 0·28$	$M_3 = 0·29$
		$M_1 = 0·33$	$M_1 = 0·32$
4. Grunerite	6·13	$M_4 = 1·96$	$M_4 = 1·97$
		$M_{123}{}^* = 4·17$	$M_{123} = 4·13$
5. Anthophyllites	1·61	$M_4 = 1·39$	
		$M_{123} = 0·22$	
	1·47		$M_4 = 1·30$
			$M_{123} = 0·17$
6. Cummingtonites	2·48	$M_4 = 1·65$	
		$M_{123} = 0·83$	
	2·50		$M_4 = 1·50^{(i)}(1·74)^{(ii)}$
			$M_{123} = 1·29\ (0·58)$

* M_{123} refers to Fe^{2+} per formula unit in the $M_1 + M_2 + M_3$ positions.
1. (a) D. Virgo and S. S. Hafner, *Amer. Mineral.* **55**, 201 (1970).
   (b) S. Ghose, *Z. Krist.*, **122**, 81 (1965).
2. C. W. Burnham, Y. Ohashi, S. S. Hafner and D. Virgo, *Amer. Mineral.* **56**, 850 (1971).
3. (a) G. M. Bancroft and R. G. Burns, *Mineral. Soc. Amer. Spec. Pap.*, **2**, 137 (1969).
   (b) J. J. Papike and J. R. Clark, *Amer. Mineral.*, **53**, 1156 (1968).
4. (a) G. M. Bancroft, R. G. Burns and A. G. Maddock, *Amer. Mineral.* **52**, 1009 (1967).
   (b) L. W. Finger, *Mineral. Soc. Amer. Spec. Pap.*, **2**, 95 (1969).
5. (a) G. M. Bancroft, R. G. Burns, A. G. Maddock and R. G. J. Strens, *Nature*, **212**, 913 (1966).
   (b) L. W. Finger, *Amer. Mineral.* **55**, 300 (1970) (abstract).
6. (a) G. M. Bancroft, R. G. Burns and A. G. Maddock, *Amer. Mineral.*, **52**, 1009 (1967).
   (b) (i) S. Ghose, *Acta Crystallogr.*, **14**, 622 (1961).
   (ii) K. F. Fischer, *Amer. Mineral.*, **51**, 814 (1966).

host of valuable data on the site preferences of cations in a variety of silicate minerals.

8.3 A survey of the method in amphiboles and pyroxenes

To illustrate the method and some of the difficulties discussed in the last two sections, it seems appropriate to look in detail at recent results on series of amphiboles and pyroxenes.

The cummingtonite-grunerite series. Room temperature spectra have been recorded for six minerals of the cummingtonite-grunerite series

Table 8.2 Comparison of Mössbauer and chemical analyses values for % Fe^{3+}/total iron

	%{Fe^{3+}/(Fe^{2+} + Fe^{3+})}	
Mineral	Mössbauer	Chemical Analyses
Howieite	24	20
Deerite	37	37
Crocidolite	41	41
Glaucophane	28	32
Crossite	37	40

G. M. Bancroft, R. G. Burns and A. J. Stone, *Geochim. Cosmochim. Acta*, **32**, 547 (1968); G. M. Bancroft and R. G. Burns, *Mineral. Soc., Amer. Spec. Pap.*, **2**, 137 (1969).

$(Fe, Mg)_7Si_8O_{22}(OH)_2$. The cation composition of these minerals is given in Table 8.3.

The crystal structure of an amphibole has been illustrated in the previous chapter (Fig. 7.13). There are four distinct positions of six-fold coordination which are designated M_1, M_2, M_3 and M_4 in the proportions $M_1:M_2:M_3:M_4 = 2/7:2/7:1/7:2/7$. The spectra consist of two doublets (Fig. 7.14) and the lines were assigned: inner doublet to Fe^{2+} in the M_4 position, and outer two lines to Fe^{2+} in the M_1, M_2 and M_3 positions. As seen in section 7.5, the M_4 site being the most distorted position, should give rise to the smallest quadrupole splitting; and the outer two lines are usually slightly broader than the inner two, suggesting that the outer two lines are due to two or more non-coincident Lorentzians. There are two other features of the peak areas which strongly indicate that this assignment is correct. First, as mentioned in section 7.3, the X-ray diffraction measurements on a cummingtonite (35·8 mole % Fe^{2+}) of very similar composition to the 35·4 mole % Fe^{2+} sample indicates that the great majority of the iron is in the M_4 position. This result may be correlated with the Mössbauer spectrum

Table 8.3 Cation composition of cummingtonites and grunerites

Fe^{2+} (mole %)	35·4	51·8	63·0	85·0	87·5	95·3	19·7
Fe^{2+} per formula unit	2·48	3·63	4·41	5·95	6·13	6·67	1·38
Mg^{2+} per formula unit	4·32	3·17	2·44	0·77	0·77	0·13	3·76
Mn^{2+} per formula unit	0·08	0·13	0·13	0·10	0·05	0·09	1·66
Ca^{2+} per formula unit	0·13	0·08	0·02	0·18	0·06	0·11	0·20

G. M. Bancroft, R. G. Burns and A. G. Maddock, *Amer. Mineral.*, **52**, 1009 (1967).

Mössbauer spectroscopy

of the 35·4% Fe^{2+} cummingtonite in which the inner (M_4) peaks are the more intense. Second, we would expect that if $A_4/A_{123} \simeq N_4/N_{123}$, the ratio of areas for a grunerite $Fe_7Si_8O_{22}(OH)_2$ would be $\frac{2}{5}$ (0·40), whereas the value obtained on the 95·3 mole % Fe^{2+} sample is 0·33 (Fig. 7.14). Thus the qualitative site population arguments reinforce the structural arguments so that there can be little doubt that the assignment is correct.

To determine the site populations in each mineral, use is made of eqn. (8.1) in the following form:

$$A_4/A_{123} = C\, N_4/N_{123}$$

where subscript 4 refers to the M_4 position and 123 to the $M_1 + M_2 + M_3$ positions. Rather than assume that $C = 1$, the data on the 95·3% grunerite provides an excellent opportunity of calculating C. We assume that all the 0·20($Mn^{2+} + Ca^{2+}$) (Table 8.3) enters the large M_4 position, and that the remaining 1·80 M_4 positions are occupied by Fe^{2+}. The large Ca^{2+} ion should certainly fill the M_4 position (as it does in actinolite), and as will be seen shortly, the Mössbauer results on the manganoan cummingtonite indicate that Mn^{2+} also prefers the M_4 position. The effect of small amounts of Mn^{2+} entering the M_1, M_2 and M_3 positions, or some Mg^{2+} entering the M_4 position, will be discussed shortly.

Applying these assumptions, and using the calculated areas (Table 8.4) for the 95·3% Fe^{2+} grunerite:

$$A_4/A_{123} = 1·91/5·74 = 0·333,\; N_4/N_{123} = 1·80/4·87 = 0·370$$
$$\text{and}\; C = 0·333/0·370 = 0·90(0)$$

We now assume that C is a constant for all the minerals in this series, and using the Mössbauer areas and chemical analyses for Fe^{2+}, the site populations in Table 8.4 can be readily calculated. These are plotted graphically in Fig. 8.1.

Before looking at the site populations more closely, it seems worthwhile to consider the following questions: first, how accurate is the value of C; second, is C a constant; third, how precise are the area ratios, and do they vary markedly with absorber thickness?

The value of C seems to be accurate to within 0·02. The maximum amount of the 0·13 Mg^{2+} that is likely to enter M_4 is 0·04. This reduces the amount of iron to 1·75 Fe^{2+} in the M_4 position and C becomes 0·93. To demonstrate the effect of this change in C on the calculated site populations for the 35·4% Fe^{2+} cummingtonite, N_4 becomes 1·64 and $N_{123} = 0·84$ compared to the previous values of 1·65 and 0·83. If on the other hand, some (0·04) of the manganese in the 95·3% Fe^{2+} grunerite is present in the M_1, M_2 or M_3

Table 8.4 Peak areas and site populations from Mössbauer spectra of cummingtonites and grunerites

Mole % Fe^{2+}	A_{123}	A_4	Fe^{2+} in M_4 per formula unit	Fe^{2+} in $M_1 + M_2 + M_3$ per formula unit	$G(X_4)\Gamma_4/G(X_{123})\Gamma_{123}$ $f_4 = f_{123} = 0.6$
35.4	1.99	3.58	1.65	0.83	0.91
51.8	1.20	0.74	1.47	2.16	0.99
63.0	2.40	1.39	1.73	2.68	1.03
85.0	3.07	1.07	1.67	4.28	1.01
87.5	2.05	0.87	1.96	4.17	1.13
95.3	5.74	1.91	1.80	4.87	1.01
19.7	1.43	0.47	0.37	1.01	—

G. M. Bancroft, R. G. Burns and A. G. Maddock, *Amer. Mineral.* **52**, 1009 (1967).

Mössbauer spectroscopy

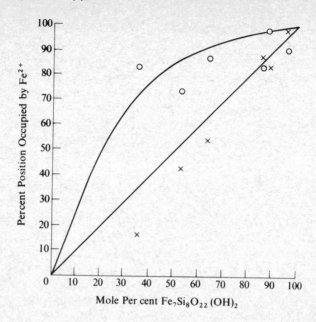

Fig. 8.1 Percentage of the M_4 and $M_1+M_2+M_3$ positions occupied by iron in the cummingtonite–grunerite series. The lines through the points for the M_4 position are those adapted from curves by Mueller (*Geochim. Cosmochim. Acta*, **26**, 581 (1960)), for a regular solid solution. $0 = M_4$ position; $X = M_1+M_2+M_3$ positions. [G. M. Bancroft, R. G. Burns, and A. G. Maddock, *Amer. Mineral.*, **52**, 1009 (1967).]

positions, then C becomes 0·88, and $N_4 = 1·66$ and $N_{123} = 0·82$. These calculations show how insensitive the results are to small errors in C arising from the choice of cation distribution in the grunerite. Obviously, however, it would be better still to use the areas from the true grunerite end member to calculate C.

The constancy of C has not yet been accurately demonstrated, but it can be readily shown that it is an approximate constant—for samples containing ≈ 10 mg cm^{-2} of Fe^{2+}—by calculating $G(X)$ values, and using the experimental Γ values. The ratios $G(X_4)\Gamma_4/G(X_{123})\Gamma_{123}$ are shown in Table 8.4, and the values indicate that this ratio does not deviate greatly from one. The rather large scatter probably reflects the large errors in estimating the $G(X)$ values, and the rather large standard deviations in the half widths. The general increase in the ratio from cummingtonite to grunerites reflects the larger M_4 peaks in cummingtonite (with $G(X_4)/G(X_{123}) < 1$), and the larger M_{123} peaks in grunerites ($G(X_4)/G(X_{123}) > 1$). It seems likely that the ratio of f values will remain nearly constant for all Fe:Mg ratios, since structural variations in the cummingtonite grunerite series are small. Thus C (eq. 8.2) should be close to 0·90 for all specimens, and this value is used to calculate the site populations in Table 8.4.

The standard deviations in the site populations are always less than ten per cent of the numbers quoted with the exception of the manganoan cummingtonite. These standard deviations could be decreased markedly with the present narrower and stronger sources. The resolution is also improved somewhat at liquid N_2 temperature, but because the peaks are well resolved at room temperature, it does not appear to be essential to take these spectra at the lower temperature.

The reproducibility of the area ratios, and the invariance with increasing absorber thickness is indicated in Table 8.5 for the 35·4% Fe^{2+} cummingtonite. We would expect the ratio to get smaller with increase in thickness because the more intense line saturates first. The variation in ratio is hardly outside the standard deviations, indicating that the saturation corrections are very small indeed (especially for absorbers having less than 10 mg cm^{-2} Fe^{2+}).

Table 8.5 Mössbauer area ratios for 35·4% Fe^{2+} cummingtonite

Fe^{2+} concentration in absorber (mg cm^{-2})	$A_4/(A_4 + A_{123})$
7·8	0·650 ± 0·010
10·0	0·643 ± 0·017
13·7	0·636 ± 0·015
23·7	0·626 ± 0·012
40·9	0·626 ± 0·026

G. M. Bancroft, *Physics Letters*, **26A**, 17 (1967).

Figure 8.1 indicates that Fe^{2+} strongly favours the M_4 position in the cummingtonite–grunerite structure over the M_1, M_2 and M_3 positions. As has been seen in Table 8.1, the Mössbauer site populations for the 87·5% grunerite and the 35·4% cummingtonite are in good agreement with the X-ray values, indicating strongly that the assumptions are justified. The site populations are scattered about the smooth curves in Fig. 8.1. This scatter is outside any error in the method, and may reflect different temperatures of formation or different cooling histories for these samples which were not derived from the same locality. The scatter of points is of course strongly influenced by the amount of Ca^{2+} and Mn^{2+} present, and the scatter between the 85·0, 87·5 and 95·3 samples is largely due to the variable (Ca + Mn) content of these samples (0·28, 0·11, and 0·20 p.f.u. respectively). Thus, the percentage of the *total* M_4 positions occupied by Fe^{2+} in the 85·0%, 87·5%, and 95·3% samples is 83·5%, 98·0%, and 90·0%; whereas the percentage of the *available* M_4 positions occupied by Fe^{2+} (assuming Ca + Mn enter the M_4 position) is 97·1%, >100% and 100%. However the very considerable disorder of the 51·8% Fe^{2+} sample cannot be attributed to the amount of Ca + Mn.

The results for the manganoan cummingtonite demonstrate the value of the Mössbauer technique for determining Mn^{2+} ordering in silicate structures. The site populations given in Table 8.4 show that Fe^{2+} does *not* order in the M_4 position in the 19·7% Fe^{2+} cummingtonite. Assuming that the 0·20 Ca^{2+} enters M_4 along with the 0·37 Fe^{2+}, then the remaining 1·43 positions must be filled by Mn^{2+}, with possibly a small amount of Mg^{2+}. It is apparent from these results that the M_4 site preference is Mn > Fe^{2+} > Mg. Recent X-ray results on another manganoan cummingtonite are consistent with the above result, although the X-ray results only show that the M_4 site was highly enriched in (Fe^{2+} + Mn) because of the similar scattering factors of Fe and Mn.

The above site populations also indicate that, even if the total 1·43 M_4 positions are filled by Mn, appreciable (0·23) Mn^{2+} enters the M_1, M_2, and M_3 positions. This result is not consistent with a previous suggestion that the maximum amount of Mn^{2+} known to enter the cummingtonite structure (about 30% Mn^{2+}) corresponds to complete filling of the M_4 positions. This result indicates that the Mössbauer technique should be valuable in estimating Fe/Mn ratios in crystallographic positions, which at the present time cannot be directly estimated by X-ray crystallography.

The orthopyroxene series. Because of the wide occurrence of orthopyroxenes $(Fe, Mg)_2Si_2O_6$ in metamorphic, plutonic and volcanic rocks, and their relatively simple structures, estimation of site populations in natural and synthetic orthopyroxenes could be very valuable as a thermal indicator. Initial results at room temperature showed that Fe^{2+} in the M_1 and M_2 sites could be resolved (see Fig. 7.8) and the two doublets in the spectra were assigned: inner doublet to the M_2 position, and outer doublet to the M_1 position, because the more distorted M_2 position should give rise to the smaller quadrupole splitting. This assignment is supported by the large inner (M_2) peaks in a hypersthene containing about 50% Fe^{2+}, and the X-ray result on an orthopyroxene of similar content which indicated that Fe^{2+} strongly ordered in the M_2 position.

For samples containing below about 50% Fe^{2+}, it is not possible to obtain precise site populations at room temperature, and Ghose and Hafner showed that the resolution of the peaks, and therefore the precision of the areas, increased markedly at 80 K. However, even at 80 K, the resolution for low Fe^{2+} orthopyroxenes is not as good as in the room temperature spectra of cummingtonite, grunerites, and anthophyllites.

Quantitative site populations in silicate minerals

To derive site populations for orthopyroxenes, we again use eq. (8.1). To determine C, it would be desirable to have Mössbauer results for the ferrosilite end member $Fe_2Si_2O_6$, for which N_2/N_1 is one. However, accurate results on $Fe_2Si_2O_6$ have not yet been reported, and naturally occurring orthopyroxenes containing greater than 85.9% Fe^{2+} are not known. As a result, most of the results to date have been reported assuming that $C = 1$. As for the cummingtonite–grunerite series, C is assumed to be a constant for the whole series. Despite the large amount of work on orthopyroxenes, there is still no direct evidence that C is a constant, although the assumption is strongly supported by the good agreement between the Mössbauer and X-ray results (Table 8.1).

Results for a large number of natural orthopyroxenes are given in Table 8.6, and a plot of Fe^{2+} in M_2 position against Fe^{2+} in M_1 is given in Fig. 8.2 for these samples. The smooth curves indicate the ideal equilibrium distribution at temperatures from 500°C to 1000°C. The origin of these curves will be discussed later. There is again an appreciable scatter of the points about smooth curves. However, the Fe^{2+} in the volcanic rocks tends to be

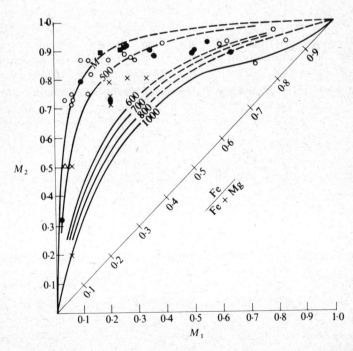

Fig. 8.2 Plot of Fe^{2+} site occupancy at M_1 against M_2 for natural orthopyroxenes. Solid lines refer to experimentally determined equilibrium isotherms at 1000°, 800°, 700°, 600° and 500°C. ○, ● = orthopyroxenes from granulite and metamorphic rocks. × = orthopyroxenes from volcanic rocks. [D. Virgo and S. S. Hafner, *Amer. Mineral.*, **55**, 210 (1970)].

Table 8.6 Site populations for natural orthopyroxenes

Samples of metamorphic rocks		Fe^{2+} Site occupancy number	
Sample	Fe/(Fe + Mg)	M_1	M_2
K-23	0.879	0.828	0.928
XYZ	0.877	0.784	0.970
X1-132	0.788	0.723	0.860
355	0.776	0.623	0.928
H-8	0.761	0.628	0.895
V-2	0.758	0.594	0.922
276	0.738	0.549	0.927
207	0.705	0.509	0.900
B1-9	0.698	0.498	0.899
7725	0.652	0.379	0.926
205	0.625	0.363	0.886
278	0.623	0.340	0.905
264	0.591	0.263	0.920
4642A	0.578	0.246	0.910
68	0.577	0.285	0.870
0-1	0.577	0.278	0.876
R1742	0.576	0.244	0.908
3209	0.574	0.251	0.897
68671	0.531	0.164	0.899
37218	0.531	0.192	0.869
7286	0.496	0.112	0.869
CH113	0.483	0.113	0.853
0-4	0.475	0.128	0.822
277	0.442	0.088	0.797
R96	0.430	0.106	0.754
37651	0.406	0.064	0.747
SP18	0.400	0.069	0.732
115/3	0.390	0.065	0.716
A	0.386	0.036	0.735
274	0.172	0.022	0.322
Samples from igneous rocks			
HK52110301	0.564	0.329	0.800
HK56071001	0.534	0.264	0.805
HK53051501	0.491	0.193	0.788
HK53051602	0.462	0.191	0.734
DOM 1	0.457	0.189	0.725
N-5	0.272	0.037	0.506
HK1955EG	0.24	0.05	0.19
HK1955KG	0.24	0.05	0.19

D. Virgo and S. S. Hafner, *Amer. Mineral.*, **55**, 210 (1970); S. S. Hafner and D. Virgo, Proc. Apollo 11 Lunar Sci. Conf., *Geochim. Cosmochim. Acta. Suppl.*, **1**, Vol. 3, p. 2183, Pergamon.

more disordered in general than in the lower temperature metamorphic rocks. Compare metamorphic and volcanic rocks of similar Fe^{2+} content. Volcanic sample HK-56071001 has 0·81 Fe^{2+} in M_2 and 0·26 in M_1, whereas metamorphic sample 68671 has 0·90 Fe^{2+} in M_2 and 0·16 Fe^{2+} in M_1. The samples from the Kii lava flow HK1955EG and HK1955KG are significantly more disordered than in other metamorphic and volcanic samples of similar composition, but the error in the site populations for these low iron samples is rather large. It is also apparent from Fig. 8.2 that the differences in ordering are not nearly as large as that expected on the basis of equilibrium distributions at the highest temperature attained.

Despite the above general differences between metamorphic and volcanic rocks, the scatter of the site populations in the metamorphic samples is often very large, especially for the samples with Fe/(Fe + Mg) less than 0·44. For these low Fe samples, the statistical errors in the areas coupled with small amounts of other cations such as Ca + Mn, could give rise to this scatter. Hafner and Virgo decided that, at present, the site populations are best represented in terms of a binary Fe, Mg solid solution because of the possibility of exsolution of the very small amount of Ca as a Ca clinopyroxene. In addition, they have made the rather dubious assumption that peak heights are proportional to peak areas, implying that the peak widths are equal. Their data, however, indicate that the widths differ by as much as $\pm 5\%$. The amount of iron in a position is proportional to the area (eq. 2.26) and not the height. Despite the above assumption, the above work, along with the heating experiments discussed in the next section, form a most valuable contribution to geochemistry.

Although the resolution of the above spectra improves markedly as the absorber temperature is lowered, the area ratios remain very similar. For specimen No. 2 in Table 8.1, the derived site populations at room and liquid N_2 temperatures differ by only about 0·01 per formula unit. When the area ratios do differ markedly with temperature, there must be more than two ferrous doublets present with markedly different temperature dependencies of the quadrupole splitting (Fig. 7.11).

8.4 A possible geothermometer

The cations in silicate structures prefer to enter (or order) in one of the several cation sites. For example, in the previous section, the Mössbauer method showed that Fe^{2+} in orthopyroxene orders in the M_2 position; while in the cummingtonite–grunerite series, Fe^{2+} orders in the M_4 position. However the extent of ordering depends on the conditions of formation (e.g., temperature and pressure), and the subsequent cooling history. Thus the cation distribution in volcanic orthopyroxene was generally more disordered than in metamorphic orthopyroxenes (Fig. 8.2).

Mössbauer spectroscopy

Using classical thermodynamics and kinetics (ref. 8.4), it is possible to indicate how the extent of ordering will vary with temperature and with the cooling history of the sample. Consider an orthopyroxene containing just Fe^{2+} and Mg^{2+} in the M_1 and M_2 positions. For the exchange reaction between cations:

$$Fe_{M_2} + Mg_{M_1} \rightleftarrows Fe_{M_1} + Mg_{M_2} \tag{8.3}$$

$$\Delta G_E = \Delta G_E^0 + RT \ln (a_{Fe_{M_1}} a_{Mg_{M_2}} / a_{Fe_{M_2}} a_{Mg_{M_1}}) \tag{8.4}$$

where: ΔG_E is the free energy change,
ΔG_E^0 is the standard free energy change at temperature T, and one atmosphere pressure, and
a's are the activities.

Two assumptions are then made. First, the activity coefficients are unity, so that $a = x$, the mole fraction. This implies that we are dealing with an ideal solution. The solution is not exactly ideal, and it is now recognized that a non-ideal model is necessary to interpret the ordering results accurately. Second, we assume that equilibrium is established before cooling occurs, and that the site populations observed reflect the highest temperature reached. This implies that no exchange (diffusion) takes place while cooling is occurring. This is not true as will be seen shortly. However, using these assumptions:

$$\Delta G_E = 0$$

and

$$\Delta G_E^0 = -RT \ln \frac{x_1(1-x_2)}{x_2(1-x_1)} = -RT \ln k \tag{8.5}$$

where x_1 and x_2 are mole fractions of Fe^{2+} in the M_1 and M_2 positions respectively. In order to obtain the expected equilibrium distributions at various temperatures and to ensure that the solution is reasonably ideal, it is necessary to determine the *equilibrium* cation distributions for: first, at least one sample at a variety of temperatures, and second, samples having different iron contents at one temperature. For one sample, the equilibrium constant (and then ΔG_E^0) was obtained (Table 8.7) at temperatures from 500°C to 1000°C. This gave one point for each temperature on the equilibrium isotherms drawn in Fig. 8.2, and the form of the curve expected from the equation,

$$k = \{x_1(1-x_2)/x_2(1-x_1)\} \tag{8.6}$$

was confirmed up to 60 mole % Fe^{2+}, by obtaining the equilibrium distributions at 1000°C for a number of samples having a range of iron content.

Table 8.7 Equilibrium constants and standard Gibbs free energy for orthopyroxenes equilibrated between 500°C and 1000°C

Temperature	Equilibrium constant k.	Standard Gibbs Free energy ΔG_E^0, kcal/mole*
500	0·051	4·57
600	0·148	3·31
700	0·173	3·39
800	0·211	3·32
1000	0·235	3·66

D. Virgo and S. S. Hafner, Mineral. Soc. Amer. Spec. Pap., **2**, 67 (1969).
* 1 cal = 4·18 joules.

The validity of eq. (8.6) was then assumed for the other temperatures. Thus the expected site populations for any equilibrium temperature are expressed by the smooth curves in Fig. 8.2. This, of course, indicates that for equilibrium the amount of ordering decreases quite markedly as the temperature is lowered. We have, then, a geothermometer of considerable potential.

Unfortunately, the observed site populations for all but the Kii orthopyroxenes (Fig. 8.2) indicated 'equilibrium' temperatures between 500°C and 600°C whereas the temperature of formation of volcanic rocks is known to be around or over 1000°C. Virgo and Hafner have found that the equilibrium reaction occurs reasonably rapidly at temperatures as low as 480°C. For example, apparent equilibrium is approached in some natural samples in 11 hours at 1000°C and after 50 hours at 500°C. Thus, unless a terrestrial pyroxene was quenched extremely rapidly from its highest equilibrium temperature, the final site populations will not directly reflect the temperature of formation. In fact, it appears as if the final site populations may reflect the rate of cooling rather than the melt temperatures.

The effective 'equilibrium' temperatures can be determined from Fig. 8.2 for any given orthopyroxene. For other pyroxenes, in which the equilibrium isotherms are not known, it is possible to determine the 'equilibrium' temperature by heating the sample over a wide range of temperatures. When disordering begins to occur, a temperature above the 'equilibrium' temperature has been reached.

It is important now to examine the kinetics of order and disorder. Considering that ΔG_E^0 is a constant over the whole temperature range, and that the ordering or disordering represented by eq. (8.3) is an ideal interchange reaction, the time rate of change of Fe^{2+} at M_2 at constant temperature is given by:

$$\frac{-dx_2}{dt} = -K_{12}x_1(1 - x_2) + K_{21}x_2(1 - x_1) \tag{8.7}$$

where K_{12} is the specific rate constant for the ordering reaction (increasing Fe^{2+} in M_2) and K_{21} is the specific rate constant for the disordering reaction (decreasing Fe^{2+} in M_2). Using the substitutions

$$x_1 = 2x - x_2$$

where

$x = Fe^{2+}/(Fe^{2+} + Mg^{2+})$ ratio in the orthopyroxene and

$$k = K_{21}/K_{12} = \frac{x_1(1 - x_2)}{x_2(1 - x_1)} \tag{8.8}$$

and integrating, the expression becomes:

$$\mp K_{21}\Delta t = (k/2a) \ln \left[\frac{\mp a - b}{a + b} \right]_{x_2'}^{x_2''} \tag{8.9}$$

where:

$$a = [k_{-1}^2(x - \tfrac{1}{2})^2 + k]^{1/2},$$

$$b = k_{-1}(x_2 - x) + \frac{k_{+1}}{2}$$

and

$$k_{-1} = 1 - k$$

$$k_{+1} = 1 + k.$$

Δt is the time interval from the start of heating or cooling, and x_2'' and x_2' are the initial and final values of x_2 after heating or cooling. The upper signs refer to the case for ordering (i.e. an increase of x_2 with time) whereas the lower signs represent disordering. From eq. (8.8), a logarithmic dependence of the site occupancy numbers with time is expected at a fixed temperature. The time dependence of x_2 was calculated for orthopyroxene 3209 (Table 8.6) (Fig. 8.3) for the disordering reaction. Thus the measured site population of Fe^{2+} in $M_2 = 0.90$ decreases to its equilibrium values given in Fig. 8.2 at different temperatures. K_{21} was calculated from Fig. 8.3 to be: $K_{21}(500°C) = 6 \times 10^{-5}$ min^{-1} and $K_{21}(1000°C) = 1 \times 10^{-2}$ min^{-1}. Using these values and the Arrhenius equation,

$$K_{21} = A \exp(-E_{aD}/RT) \tag{8.10}$$

the activation energies E_{aD} for the disordering process can be calculated to be 20 Kcal per mole. From eqs. (8.7) and (8.8), the activation energy for the ordering reaction E_{ao} can be calculated to be 16·5 kcal per mole.*

The low activation energies for both ordering and disordering processes initially indicate that the observed site populations in terrestrial or lunar

---

* 1 cal = 4.18 joules.

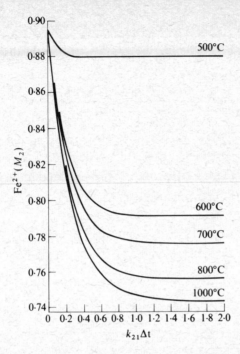

Fig. 8.3 Computed site occupancy number x_2 for sample 3209 (Table 8.6) as a function of time Δt on the basis of the ideal distribution model: disordering from, D. Virgo and S. S. Hafner, *Min. Soc. Amer. Spec. Pap.*, **2**, 67 (1969).

samples will not reflect the crystallization temperatures, but will strongly reflect the rate of cooling. On geological time scales, it appeared that all natural specimens would give site populations close to the temperature ($\approx 480°C$) at which the ions no longer diffuse (Fig. 8.2). However more recent results strongly indicate that the kinetics of ordering occurs in two steps, governed by two rate constants K'_{12} and K''_{12}. The first ordering step from $1000° - \approx 600°C$ is at least qualitatively consistent with the ideal distribution model, and is very rapid (short range). The second step, from about $600°C$ to $480°C$ in orthopyroxenes is very slow (long range), and K''_{12} is close to zero over a period of months at $550°C$. The ordering process in this temperature range could quite likely take place on a geological time scale.

In agreement with the above ideas, Virgo and Hafner took an orthopyroxene (3209, Table 8.6) having $Fe^{2+}_{M_1} = 0.25$ and $Fe^{2+}_{M_2} = 0.90$. They partially disordered the site populations at $900°C$ ($Fe^{2+}_{M_1} = 0.41$, $Fe^{2+}_{M_2} = 0.73$) to simulate melt conditions, and then heated the pyroxene at $550°C$ for weeks. The Fe^{2+} distribution reached a constant value ($M_1 = 0.37$;

Mössbauer spectroscopy

$M_2 = 0.78$) after about 47 hours and did not change for three months. Note that compared to the initial cation distribution, the Fe^{2+} is still considerably disordered after the heating at 550°C. The remainder of the ordering must have taken place over a very long period of time. The results on a lunar pigeonite described shortly are consistent with this two step mechanism.

The lunar samples returned from recent Apollo 11 and 12 flights have provided interesting pyroxene and olivine specimens for order–disorder studies. Unfortunately, amongst the pyroxene specimens, only one pigeonite sample from an Apollo 12 rock was reasonably homogeneous. All other separates had wide ranges of composition—usually on a short-range scale. This inhomogeneity gives rise to broad Mössbauer lines, and the area ratios are temperature dependent.

The assignment of the Fe^{3+}-free lunar clinopyroxenes is as given previously: inner doublet to Fe^{2+} in M_2, and outer doublet to Fe^{2+} in M_1. Average site populations for some Apollo 11 separates are given in Table 8.8, along with average compositions. These samples all show highly ordered

Table 8.8 Site populations for Fe^{2+} in lunar clinopyroxenes

	'Average' composition			Site populations					
	Atoms per formula unit			M_1			M_2		
	Ca	Fe	Mg	Ca	Fe	Mg	Ca	Fe	Mg
1.	0.64	0.54	0.82	0	0.18	0.82	0.64	0.36	0
2.	0.36	1.14	0.50	0	0.56	0.44	0.36	0.58	0.06
3.	0.56	0.54	0.88	0	0.18	0.82	0.56	0.36	0.08
4.	0.72	0.40	0.88	0	0.14	0.86	0.72	0.26	0.02
4'	4 heated at 680°C			0	0.15	0.85	0.72	0.25	0.03
4"	4 heated at 1000°C			0	0.17	0.83	0.72	0.23	0.05

P. Gay, G. M. Bancroft and M. G. Bown, *Proc. Apollo 11 Lunar Sci. Conf., Geochim. Cosmochim. Acta Suppl. 1*, vol. 1, Pergamon (1970); S. S. Hafner and D. Virgo, *Proc. Apollo 11 Lunar Sci. Conf., Geochim. Cosmochim. Acta Suppl. 1*, vol. 3, Pergamon (1970).

Mg and Fe^{2+} distributions which are indicative of fairly low 'equilibrium' temperatures. Heating sample 4 above 680°C shows that the sample begins to disorder, so that the 'equilibrium' temperature must be lower than 680°C. However, as seen from Table 8.8, the changes in the site populations after heating are very small.

More extensive studies on Apollo 12 pyroxenes indicate that the free energy of the exchange reaction increases with rising Ca content of the pyroxenes i.e., from ≈ 3.6 kcal (Table 8.7) for orthopyroxenes to 4.0 kcal for pigeonite and 7–10 kcals for sub-calcic augites. Order–disorder heating experiments on a pigeonite similar to those carried out for sample 3209 shows that the pigeonite cooled rapidly to $\approx 810°C$, and then cooled slowly over the temperature range 810–480°C.

8.5 Site populations and crystal field phenomena

As demonstrated in the previous sections of this chapter, cations tend to order in one or more structural positions in silicates. Although we have indicated that large cations tend to order in the large cation sites (e.g. Ca^{2+} in the pyroxene M_2 position) and small cations in the small cation sites (e.g. Mg in the pyroxene M_1 position), and free energies for the exchange reaction (eq. 8.3) are known for some pyroxenes, it would be desirable to be able to predict ΔG_E (and thus the extent of ordering) for exchange reactions such as eq. (8.3) by calculating the total energy of cations in the different silicate positions. Considering silicates to be ionic, this total energy will include such terms as the lattice energy, crystal field stabilization energy (CFSE) for transition metal ions, and entropy contributions. Or on a covalent picture of silicates, the total energy would include bond energies instead of lattice energies. For such complex structures such as silicates, the calculation of the larger energy contributions—lattice energies or bond energies—seems insuperably difficult without many questionable assumptions. In his excellent book, Burns (ref. 8.5) chose to use the difference in CFSE as a guide to the sign of ΔG_E to explain ordering of Fe^{2+} in silicate structures. The CFSE could be derived from electronic spectra in conjunction with site population data from Mössbauer and X-ray studies.

Before considering the ordering of Fe^{2+} in silicates, the importance of CFSE is illustrated by the site preference problem in the mixed metal oxides which have the spinel (or inverse spinel) structures. The general formula of a spinel is given by $ABCO_4$, where A, B and C are cations whose total charge is $+8$. The spinel structure may be described as a cubic close packed array of oxide ions with one third of the cations occupying tetrahedral holes, while the remaining two thirds occupy octahedral holes. In the normal spinel structure, the A ions have a $+2$ charge, and occupy the tetrahedral sites, while the B and C ions have $+3$ charges and occupy octahedral sites. However, in the inverse spinel structures, the divalent ions (A) occupy one half the octahedral sites, while the trivalent ions B and C occupy the other half of the octahedral sites and tetrahedral sites respectively. Crystal field stabilization energies can be used to explain the inversion in all cases. We neglect tetragonal distortions in the following treatment. For example, Fe_3O_4 is inverted while Mn_3O_4 and Co_3O_4 are normal. All ions here are high spin except for Co^{3+} which is low spin. The d level diagrams for the ions considered here, along with the CFSE energies (in terms of Δ_o or Δ_t) for each electronic configuration are given in Fig. 8.4. The approximate values of Δ_o for each configuration are also given in Fig. 8.4.

Mössbauer spectroscopy

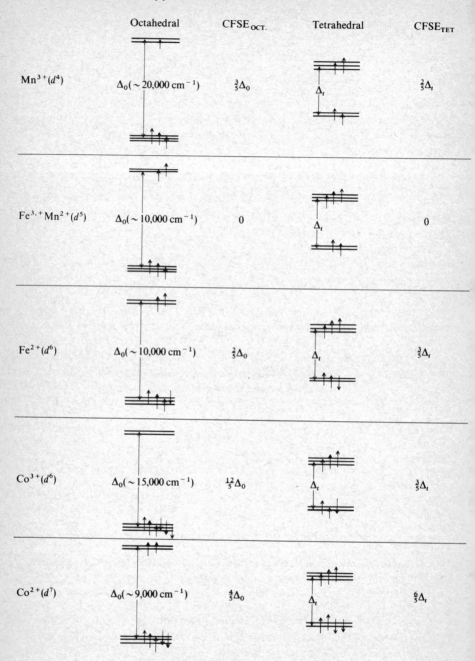

Fig. 8.4 The ground state d orbital occupancy schemes for some octahedral and tetrahedral transition metal complexes. Included are CFSE energy values in terms of crystal field splittings Δ_o or Δ_t.

For Co_3O_4, consider the exchange reaction

$$Co^{2+}_{(tet)} + 2 Co^{3+}_{(oct)} \rightleftarrows Co^{2+}_{(oct)} + Co^{3+}_{(oct)} + Co^{3+}_{(tet)}$$

$$\text{normal} \rightleftarrows \text{inverted} \qquad (8.11)$$

The overall ΔH^0 for this reaction will include, amongst other terms, the difference in lattice energies as well as the differences in CFSE's. Assuming that the CFSE's are most important in determining at least the sign of ΔH^0, then for the above reaction

$$\Delta H_{CFSE} = -\tfrac{3}{5}\Delta_t - \tfrac{4}{5}\Delta_o + \tfrac{12}{5}\Delta_o + \tfrac{6}{5}\Delta_t$$

Since $\Delta_t \sim \tfrac{1}{2}\Delta_o$, then $\Delta H_{CFSE} \sim 30\,000$ cm^{-1} or $\sim +86$ kcal/mole. Thus the above equilibrium should be far to the left (neglecting entropy terms), and the normal spinel should be formed.

In contrast, Fe^{3+} shows no CFSE in either tetrahedral or octahedral coordination (Fig. 8.4), and the ΔH_{CFSE} for the iron spinel is given by

$$\Delta H_{CFSE} = -\tfrac{2}{5}\Delta_o + \tfrac{3}{5}\Delta_t \simeq -\tfrac{1}{10}\Delta_o \sim -3 \text{ kcal/mole}$$

Thus the inverted structure would be favoured. For Mn_3O_4, ΔH_{CFSE} is positive and Mn_3O_4 is a normal spinel as expected from the positive ΔH_{CFSE}.

Table 8.9 Fe^{2+} crystal field stabilization energies and observed and predicted ordering in silicates

Mineral	Position	CFSE (kcal/mole*)	Predicted enrichment	Measured enrichment
Olivine	M_1	12.9–12.0	$M_2 > M_1$	$M_2 > M_1$
	M_2	13.1–12.8		
Orthopyroxenes	M_1	11.5–11.0	$M_2 > M_1$	$M_2 > M_1$
	M_2	11.7–11.3		
Cummingtonite	M_4	11.2–10.9	$M_4 > M_{123}$	$M_4 > M_{123}$
	M_{123}	11.0–10.7		

* 1 cal = 4.18 joules.
R. G. Burns, *Mineralogical Applications of Crystal Field Theory*, Cambridge University Press (1970).

Burns applied the same approach to the more complex silicate structures. From the polarized electronic spectra of silicates, the CFSE's for Fe^{2+} in different silicate positions were calculated. (Table 8.9). Table 8.9 shows that the CFSE for Fe^{2+} in the M_2 position in orthopyroxenes is larger than that for Fe^{2+} in the M_1 position; similarly for the cummingtonite–grunerite series, the CFSE for Fe^{2+} in M_4 is larger than that for

Mössbauer spectroscopy

Fe^{2+} in M_{123}. Recalling the ordering of Fe^{2+} in these series (Table 8.1), it is apparent that the correct ordering scheme is predicted from the CFSE's. However there are several problems which may prevent such CFSE's being widely used to explain or predict site preferences. First, silicate electronic spectra are usually very complex, and it is often difficult both to assign peak positions accurately, and assign peaks to particular $d-d$ transitions. Second, the so called centroid rule has to be assumed when calculating CFSE from the $d-d$ transitions, and this is known not to hold for simpler structures. Third, the differences in CFSE between Fe^{2+} in different positions (Table 8.9) is extremely small (~ 0.2 kcal/mole), in contrast to the very much larger values for spinels noted earlier. It would seem likely that the differences in other energy terms such as lattice energies would be much larger than this. In addition, the extent of Fe^{2+} ordering in M_2 (calculated using Boltzmann's Law and ΔCFSE = 0.2 kcal) is extremely small. Although the above approach is not an entirely satisfactory solution to a very complex problem, it seems likely that a more accurate and practical treatment will be a long time in coming.

Problems

1. Using the chemical compositions and the experimentally determined areas in Table 8.4, calculate both the Fe^{2+} and Mg^{2+} site populations in the cummingtonite–grunerite series (Table 8.4, column 5). Take $C = 0.90$, and assume that only Fe^{2+} and Mg^{2+} are present in the seven cation positions.
2. Ordering of cations has now been observed in a large number of silicate minerals. However, there has been considerable debate about ordering in the M_1 and M_2 positions in ferromagnesium olivines $(Fe, Mg)_2 SiO_4$. For example, electronic spectra (R. G. Burns, *Acta Cryst.*, A**25**, 5 (1969)) indicated a slight enrichment of Fe^{2+} ions in the M_2 positions for most Fe–Mg olivines. However, Bush *et al.* (W. R. Bush, S. S. Hafner and D. Virgo, *Nature*, **227**, 1339 (1970)) suggested that because of the quoted relatively large errors, Burns' 'statement cannot be considered significant.' Bush *et al.* obtained the Mössbauer spectra of several olivines at high temperatures and claimed to observe ordering of Fe^{2+}. Critically evaluate this claim after reading the above paper by answering the following questions. (a) Why are the widths of peaks B_1 and B_2 different? Remembering that area $\propto$ width $\times$ height, calculate height ratios H_{B_1}/H_{B_2}. (b) Why have area ratios been used in this paper rather than height ratios (or intensity ratios) in previous papers? (c) What other assumptions are made? Are the ordering conclusions significant?
3. Triplite has the general formula $(Mn, Fe, Mg)_2 PO_4 F$, and an X-ray study indicates that there are two equally populated metal sites M_1 and M_2.

The X-ray study of one specimen of composition

$$(Mn_{0.95}Fe_{0.25}Mg_{0.7}Ca_{0.1})PO_4F$$

indicates the following site populations:

$$M_1 : 0.717 \text{ Mn, Fe} + 0.231 \text{ Mg} + 0.052 \text{ Ca}$$

$$M_2 : 0.483 \text{ Mn, Fe} + 0.469 \text{ Mg} + 0.048 \text{ Ca}$$

Comment on the potential use of Mössbauer spectra for determining the complete cation ordering picture in triplite. Using the Mössbauer spectra in E. Kostiner, *American Mineralogist* **57**, 1109 (1972), attempt to assign the peaks to the two metal sites, and comment on the ordering of Fe.

4. A clinopyroxene sample yielded a Mössbauer spectrum which could be resolved into two quadrupole split doublets with Mössbauer parameters of (a) C.S. = 1.26 mm s^{-1}, Q.S. = 2.28 mm s^{-1} and (b) C.S. = 1.23 mm s^{-1}, Q.S. = 1.88 mm s^{-1}. The ratio of the intensity of the doublets a:b is $0.601:1.0$. Electron probe microanalysis of the clinopyroxene yielded the following composition expressed in terms of component oxides MgO 28.37, Al$_2$O$_3$ 0.86, SiO$_2$ 55.65, CaO 2.32, TiO$_2$ 0.29, Cr$_2$O$_3$ 0.77, FeO 13.37 wt% (total = 101.63 wt%). From a knowledge of the crystal chemistry of pyroxenes and the above information, suggest what the populations of the different cation sites might be and the oxidation states of the cations. Express your answer in terms of a formula of the type $(X_a^+ Y_b^{2+} Z_c^{3+})(M_d^{3+} N_e^{4+})(O_f^{2-})$ etc., where $a, b, c \ldots$ are the number of ions (or fractions thereof) per formula unit (expressed to two decimal places).

References

8.1 Papers in: *Mineral. Soc. Amer. Spec. Pap.* **2** (1969).
8.2 R. G. Burns and R. G. J. Strens, *Science*, **153**, 890 (1966).
8.3 G. M. Bancroft and R. G. Burns, 5th Int. Mineral Assoc. Meeting, Cambridge 1966, *Mineral Soc. Spec. Paper*, 36 (1968), The Mineralogical Society.
8.4 D. Virgo and S. S. Hafner, *Mineral Soc. Amer. Spec. Pap.* **2**, 67 (1969).
8.5 R. G. Burns, *Mineralogical Applications of Crystal Field Theory*, Cambridge University Press (1970).

Bibliography

R. G. Burns, *Mineralogical Applications of Crystal Field Theory*, Cambridge University Press (1970).
F. A. Cotton and G. Wilkinson, *Advanced Inorganic Chemistry*, Wiley Interscience (1972).

9. The Mössbauer spectra of multi-phase assemblages

As seen in chapter 7, the Mössbauer spectrum of an iron-containing mineral is characteristic of that mineral, and is usually significantly different from spectra of other minerals. In the last chapter, we discussed the use of Mössbauer areas to determine the amount of iron in structurally different positions in one mineral phase. The availability of a large number of Mössbauer spectra of minerals would suggest the following two applications for the study of bulk rock samples. First, the Mössbauer spectra should be useful in identifying the iron-containing minerals using the fingerprint technique. Second, from the relative areas of the peaks, it should be possible to obtain accurate estimates of the relative numbers of iron atoms in each mineral phase, and an estimate of the bulk ferric to ferrous ratio. If the chemical composition of each mineral phase and the weight per cent of non-iron minerals are known from chemical analysis, then accurate weight percentages of all the minerals in a bulk sample could be obtained.

Before discussing the method and results which illustrate the above two points, it seems worthwhile to comment on the apparent advantages and disadvantages of the Mössbauer technique over other methods. A Mössbauer spectrum of a bulk sample generally enables identification of many of the iron containing minerals. However, in many cases, optical examination or X-ray methods provide a more rapid and accurate identification of the component minerals. The Mössbauer effect provides substantial advantages over optical and/or X-ray techniques for poorly crystallized materials or for particles of very small grain size. For example, sharp X-ray patterns are observed for crystallites from 100 000 to 2000 Å in diameter, but diffuse lines are observed for crystallites from 2000 to 100 Å, and no observable lines are seen below 100 Å. Mössbauer spectra on the other hand can be seen for any crystallite size, although effects such as superparamagnetism become important for small crystallites. A good example of the relative utility of X-ray and Mössbauer methods is given by the attempted identification of the iron phase in fine grained manganese modules (ref. 9.1). X-ray techniques were not useful in identifying the state of the iron, but

Mössbauer spectra at 77 K gave a single doublet (C.S. = 0·60 mm s^{-1} and Q.S. = 0·78 mm s^{-1}) characteristic of Fe^{3+}. Because of the small size of the particles, a gradual transition to a six-line spectrum takes place at lower temperatures. These spectra could be attributed to α-FeOOH or γ-FeOOH, although there could be a mixture of both. Other combinations of ferric oxides, or a mixed iron–manganese oxide are also possibilities.

There are however several disadvantages which make it imperative that Mössbauer be used in conjunction with other techniques for mineral identification. The Mössbauer effect suffers from the same disadvantage as any spectroscopic technique (including X-ray powder patterns): one identifies a mineral by comparison of the spectrum with known standard spectra. There may be several strongly overlapping Lorentzians from iron in a number of minerals. This overlap makes it difficult, if not impossible to resolve the component peaks unless information on the mineral content is available from other techniques. Many of the silicate spectra shown in the last two chapters would, if superimposed, give closely overlapping peaks. For example, the similarity of Mössbauer parameters for Fe^{3+} ions in silicates makes Mössbauer spectroscopy a relatively poor technique for identifying ferric paramagnetic minerals. And for two pyroxene assemblages, or for pyroxene/amphibole assemblages, the similarity of Mössbauer parameters for most pyroxenes and amphiboles makes Mössbauer a relatively poor technique for identification. For some ferro- or antiferromagnetic minerals such as FeOOH, the expected six-line magnetic pattern is not observed if the particles are very small (~ 100 Å). It is important then to take spectra at a range of temperatures: at low temperatures a magnetic spectrum is observed, while at high temperatures, a simple quadrupole doublet may be observed. Because of these disadvantages, Mössbauer spectra will probably be of limited use for identification purposes.

The Mössbauer effect is probably of much greater importance for obtaining quantitative or semi-quantitative estimates of relative amounts of some iron-containing minerals and for analysis of the ferric to ferrous ratio in a rock sample. For the relative amounts of iron-containing minerals, there are a number of assumptions in the method, and the derived numbers are probably only of semiquantitative significance. These assumptions will be discussed shortly. However, even semiquantitative analysis seems to be often more reliable than that obtained from optical investigations, X-ray powder photographs, or the norms from chemical analysis. For example, when iron metal and iron sulphides are present in a basically silicate bulk sample, it is normal to analyse for total iron, then iron as sulphide and iron as iron metal. The iron in silicates is then calculated by difference. There are a number of difficulties here, especially in the iron metal analyses, which make the derived percentages rather inaccurate.

Mössbauer spectroscopy

In the previous chapter, we discussed the accurate determination of ferric to ferrous ratios from the Mössbauer areas. Even in bulk spectra, the ferric peaks are usually well resolved from the ferrous peaks, and the ferric to ferrous ratio can be readily obtained. This ratio is often difficult to obtain by chemical analysis, especially if the sample is difficult to dissolve or other elements such as Ti, which can exist in a number of oxidation states, are present. In either case, oxidation or reduction of the iron can take place during the analysis.

Very little work has been done on bulk samples. Sprenkel–Segel and Hanna (ref. 9.2) first used the Mössbauer technique for analysing stone-meteorites and Herzenberg (ref. 9.3) first looked at a number of terrestrial rock samples. The recent lunar flights have provided a great impetus for work on bulk samples, and provide good examples of the usefulness and difficulties of the Mössbauer technique in this area of research.

9.1 The method and general assumptions

The first step in the analyses is to assign each line to a particular mineral, and compute the area under each peak. If the lines are closely overlapping, the above analyses may be rather difficult, and the computed areas may have a rather large error. It is usually necessary to assume that the areas are directly proportional to the number of iron atoms in each mineral. This assumption implies that: first, every mineral contains the same percentage of iron as ^{57}Fe, i.e. that there is no isotope fractionation between sites; and second, that the Mössbauer f factors are identical for each iron atom in every mineral. In addition, we assume that the component peaks of a quadrupole doublet have equal intensity, and the component peaks of the magnetic peaks have the ratio $3:2:1:1:2:3$. The inner magnetic peaks are usually swamped by the paramagnetic components, and it is usual to obtain the total intensity for a magnetic mineral by multiplying the outer two peak intensities by two. The above assumptions are reasonable, but will lead to possible errors in the derived numbers.

It is usually desirable to express the mineral contents as weight per cent of the total sample to make the numbers easily comparable with those obtained from chemical analyses. This can be easily done if the weight per cent of iron in each mineral is known, and if the weight percentage of the non-iron components are known. Usually, a reasonably accurate value of the latter can be had from the C.I.P.W. norms from the total chemical analysis. For example, the total titanium content is expressed in conjunction with the appropriate amount of iron as ilmenite. However, because titanium is usually present in other minerals, the C.I.P.W. norm value for ilmenite will usually be larger than the actual amount of ilmenite in the mode.

As a simple example of the above procedure, consider the following hypothetical results (Table 9.1). A rock sample contains three minerals,

The Mössbauer spectra of multi-phase assemblages

Table 9.1 Weight percentages of iron-containing minerals from Mössbauer areas

	Mineral Y	Mineral Z	Non-iron* Minerals
Mössbauer area (% of total)	40	60	—
Weight % Fe*	10	20	—
Weight % of sample	40	30	30

* From chemical analyses.

Y and Z which contain iron, and a non-iron containing mineral which makes up 30% of the total weight. The Mössbauer areas show that 40% of the total area is due to mineral Y, and 60% of the area is due to mineral Z. Chemical analyses indicate that mineral Y contains 10 wgt % Fe^{2+} and mineral Z, 20 wgt % Fe^{2+}. The relative weight percentages of minerals Y and Z, $W_Y:W_Z$, is given by $(10 \times 40):(5 \times 60) = 40:30$. Since minerals Y and Z make up 70 wgt % of the sample, the percentages of Y and Z of the total sample are 40% and 30% respectively.

9.2 Meteorites

Meteorites have long interested geochemists in their attempt to reconstruct the early history of the solar system. Chondrites, by far the most abundant type of meteorites, are characterized by the presence of chondrules, generally spherical bodies up to a few millimeters in diameter which are composed of silicate, metal and sulphide phases. The chondrules are embedded in a matrix that varies in amount and composition, although the matrix generally has a similar content to the chondrules, but contains most of the metal and troilite. The 'ordinary' chondrites are subdivided (on the basis of the amount of iron in the metal and silicate phases) into the bronzite, hypersthene, and amphoterite groups.

Surprisingly, the olivines and pyroxenes in the ordinary chondrites are of remarkably constant composition both in the chondrules and the matrix material. The uniformity in silicate composition is thought to result from an equilibrium distribution of positive ions at an elevated temperature. As the temperature fell, decreased ion mobility caused the distribution representative of about 800°C to remain frozen in. For this reason, a chondrite exhibiting uniform olivine and pyroxene compositions is described as equilibrated.

Since all the meteorites of a given 'ordinary' chondrite group have olivines and pyroxenes of similar composition, the ferromagnesium silicate material must have been produced by one of two mechanisms; the material was derived from a single reservoir that was a relatively homogeneous mixture

Mössbauer spectroscopy

of olivine or pyroxene, or the ferromagnesium material accumulated in random proportions from separate olivine and pyroxene reservoirs. The Mössbauer study described in the next few paragraphs should illustrate the general method for calculating percentages of minerals, and indicate strongly that the first mechanism was involved.

The spectra of four meteorite samples are shown in Fig. 9.1. The inner lines at 0·3 and 2·5 mm s^{-1} are due to Fe^{2+} in the orthopyroxene M_2 position; the lines at -0.2 and 3·0 mm s^{-1} are due to absorption by Fe^{2+} in olivine. The smaller lines at high velocities are due to absorption by troilite (FeS)

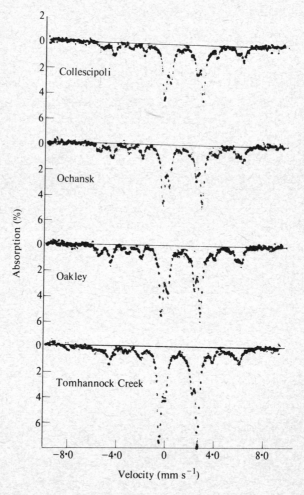

Fig. 9.1 Mössbauer spectra of four bronzite chondrites. [E. L. Sprenkel–Segel and G. J. Perlow, *Icarus* **8**, 66 (1968).]

and kamacite (Fe, Ni alloy). By inspection, it is readily seen that the ratio of the olivine area to the pyroxene area is reasonably constant. To obtain semi-quantitative results for the olivine to pyroxene ratio, the following procedure was used. From the Mössbauer spectra of chemically analysed olivines and pyroxenes, it was found that one iron atom in olivine gives rise to 1·17 times the absorption of one iron atom in orthopyroxene i.e.

$$f(\text{olivine})/f(\text{pyroxene}) = 1 \cdot 17 \pm 0 \cdot 06.$$

To calculate the ratio of iron atoms in olivine and pyroxene from the area ratio, the usual formula is used:

$$A_o/A_p \propto f_o/f_p \cdot \text{Fe}_o/\text{Fe}_p. \tag{9.1}$$

Thus the iron atom ratio (column 2, Table 9.2) is calculated by dividing the area ratio by 1·17. From microprobe analyses, the atom fraction $x = \text{Fe}/(\text{Fe} + \text{Mg})$ was obtained, and the ratio of the number of formula units of pyroxene and olivine can then be calculated from:

$$N_p/N_o = 2(\text{Fe}_p/x_p)/(\text{Fe}_o/x_o) \tag{9.2}$$

where N_p is the number of pyroxene (Mg, Fe)SiO$_3$ formula units,
N_o is the number of olivine (Mg, Fe)$_2$SiO$_4$ formula units.
The data are summarized in Table 9.2. A slight correction to N_p/N_o ($\sim$0·03) is made because of small amounts of Ca present in the pyroxene.

Table 9.2 Relative abundances of olivine and pyroxene in bronzite chondrites

Chondrite	A_o/A_p^1	$\text{Fe}_o/\text{Fe}_p^2$	x_o/x_p^3	N_p/N_o^4
Collescipoli	2.01	1.71	1.09	1.30
Ochansk 1	2.00	1.71	1.12	1.33
Ochansk 2	1.85	1.59	1.12	1.44
Oakley	1.82	1.56	1.09	1.44
Tomhannock	2.43	2.08	1.10	1.08

[1] From Mössbauer areas.
[2] Using eq. (9.1).
[3] From microprobe analyses.
[4] Using eq. (9.2).
See: E. L. Sprenkel-Segel and G. J. Perlow, *Icarus*, **8**, 66 (1968).

The uniformity in the olivine to pyroxene ratio, taken with the afore-mentioned uniformity in composition of the pyroxene and olivine, supports the conclusion that the bronzite chondrites in this study were derived from a

Mössbauer spectroscopy

common ferromagnesium silicate reservoir. If they formed from two separate reservoirs, one would expect a large variation in N_p/N_o. This study cannot indicate whether equilibration of the pyroxene and olivine took place before or after the accretion of the reservoir material to form the parent body.

It is worth mentioning some of the difficulties and assumptions in order to obtain the N_p/N_o values. Sprenkel–Segel and Perlow assumed that there was no Fe^{2+} in the M_1 orthopyroxene position. These lines would strongly overlap the olivine lines, but in low Fe orthopyroxenes only about five per cent of the iron would be in this position. In addition there is a small

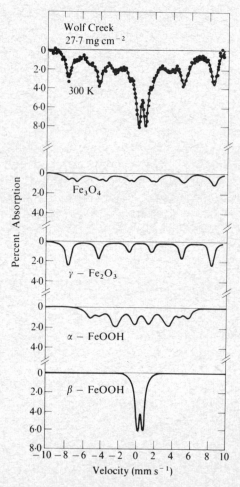

Fig. 9.2 Mössbauer spectrum of the completely oxidized Wolf Creek meteorite at room temperature, and the components resulting from computer analysis. (E. L. Sprenkel–Segel, *J. Geophys. Research*, **75**, 6618 (1970)).

amount of diopside in the sample which again would overlap the olivine positions. There are also the usual statistical errors due to the strong overlap of lines, and uncertainties due to the subtraction of the very small troilite or iron lines from the pyroxene or olivine intensities. Sprenkel–Segel and Hanna concluded that the ratio A_o/A_p carries an error of about 16%, but the relative values should be much more accurate than this, since most of the errors should be in the same direction.

Despite the rather large error, it seems likely that these numbers are more objective and accurate than those obtained by previous methods: optical examination, X-ray powder photographs and normative composition from chemical analyses. Craig calculated an average N_p/N_o of 1·28 for the equilibrated bronzite chondrites from chemical analysis, but concluded that the chemical data are not precise enough to determine the ratio in an individual bronzite chondrite.

Mössbauer spectra of oxidized meteorites show the advantages of Mössbauer over other techniques both for mineral identification purposes and quantitative estimation of mineral phases. Oxidation of meteorites occurs during meteorite fall and, of course, is a result of terrestrial weathering processes. The oxidation products usually exist as fine grained mixtures which are difficult to separate mechanically, or identify optically. The most useful technique for mineral identification is X-ray diffraction, but X-ray results can often be ambiguous because of strongly overlapping patterns, or weak and diffuse patterns from very small crystallites.

The room temperature Mössbauer spectrum of a completely oxidized iron meteorite is shown in Figure 9.2, along with the spectra of the proposed constituents. Because of the small amounts of Fe_3O_4 and α-FeOOH, these assignments remain questionable, but spectra at lower temperatures are consistent with these assignments. Previous X-ray results gave no indication of β-FeOOH or Fe_3O_4, presumably because of the very small particle size of β-FeOOH or the very small amounts of Fe_3O_4. The quantitative evaluation of this spectrum yields the following area percentages, and weight percentages. As usual, the f factors for each oxide were assumed to be identical.

Compound	γ-Fe_2O_3	Fe_3O_4	α-FeOOH	β-FeOOH
Area (%)	27	16	38	19
Molecular proportions (%)	17	7	50	26

Other spectra of oxidized chondrite fragments in achondrites give intense lines characteristic of β-FeOOH, even though several recent chemical analyses reported no Fe^{3+}. Since iron is often apportioned to different

Mössbauer spectroscopy

mineral species such as Fe, and FeS on the basis of chemical analyses, it is obviously extremely important to know the Fe^{3+} content to correctly assign the iron to the various iron-containing minerals.

9.3 Lunar soils and rocks

In the attempt to unravel the origin and history of the moon, many diverse techniques have been used to study the lunar samples from Apollo 11 and 12. Several workers have used the Mössbauer effect to indicate qualitatively the overall composition of soil and rocks, while others have used the technique to obtain semiquantitative modal percentages for the individual minerals. In the next few pages we will be mainly concerned with the latter applications.

The spectrum of a lunar soil (10084) from Tranquility Base are shown in Fig. 9.3a. The assignment of the peaks is as follows: peaks 1 and 1' to Fe^{2+} in ilmenite, peaks 2,2' to Fe^{2+} in the pyroxene M_2 positions and Fe^{2+} in glasses, peaks 3 and 3' to Fe^{2+} in pyroxene M_1 positions, peaks 4 and 4' to Fe^{2+} in olivine, 5 and 5' to an overlap of peaks 2, 3 and 4; 6, 6' and 6" to iron metal, 7 and 7' to troilite and 8, 8' and 8" to a magnetic iron spinel. The other magnetic peaks are buried under the paramagnetic peaks. Although the relative intensities of these peaks change from sample to sample, the above minerals are present in most samples.

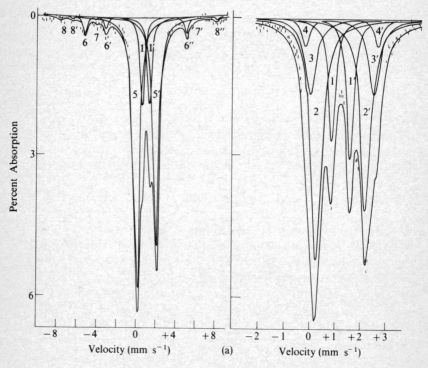

(a)

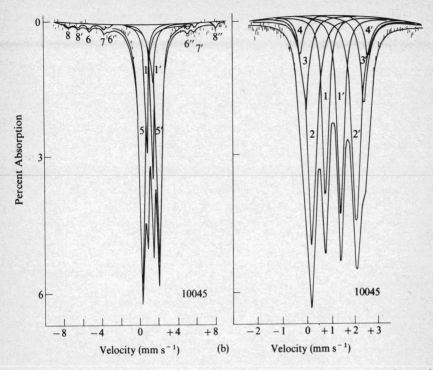

Fig. 9.3 Mössbauer spectra of lunar soils and rocks. (a) Apollo 11 lunar soil at low and high velocity. (b) Appollo 11 basalt at high and low velocity. (P. Gay, G. M. Bancroft, and M. G. Bown, *Proc. Apollo 11 Lunar Sci. Conf. Geochim. Cosmochim. Acta Suppl.* 1, vol. 1, p. 481, Pergamon.)

There are considerable computational difficulties with spectra as complex as this, and it seems worthwhile to indicate the constraints that are necessary to obtain a fit. First, the olivine peak positions were constrained to be at the values obtained from a separated olivine from a lunar rock. Second, the widths of peaks 2, 2′, 3 and 3′ were constrained to be equal but at no fixed value; the intensities of 3 and 3′ were also constrained to be equal. Third, in the high velocity spectra, the peak widths of the small peaks (6, 6′, 6″, 7, 7′, 8, 8′ and 8″) were constrained to be equal but at no fixed value.

The areas derived from the above fits are considered to be only of semi-quantitative significance for the following reasons. First, the pyroxene has a very wide composition range, and there may be appreciable quantities of glass, pigeonite, and a pyroxenoid structure present, all of which have slightly different quadrupole splittings. Each pyroxene peak then is an average of a large number of closely overlapping Lorentzians and will not be exactly Lorentzian. The relative pyroxene and olivine peak intensities will be affected by the degree of deviation from Lorentzian shape. Second, although the olivine composition appears to be reasonably constant in the lunar samples, the olivine Q.S. is known to vary with composition and any small change in the olivine Q.S. could make a significant change in the calculated olivine intensity.

The parameters obtained for all the minerals are internally and externally consistent: i.e., they agree from sample to sample, and agree with separated minerals of similar composition. Peaks 8, 8' and 8" deserve special comment, as the interpretation of these peaks indicates the need to use Mössbauer in close conjunction with other techniques. These peaks were initially assigned to magnetite, because their positions agreed with the known positions of Fe_3O_4. However, other workers using a variety of techniques were unable to find magnetite, and there is not any evidence for an appreciable concentration of Fe^{3+}. An iron, titanium, chromium spinel has however been found in the dust and rocks, and since it seems likely that this would be magnetic, these peaks could be due to this spinel, although the possibility of these peaks being due to Fe_3O_4 from terrestrial oxidation of Fe cannot be ruled out. These very small peaks cannot be readily seen in the spectrum, but their existence is substantiated by considerable statistical evidence.

The spectrum of a crystalline basalt is also shown in Fig. 9.2b. The percentages of the total Mössbauer area and the weight per cent of total sample for each mineral are summarized in Table 9.3 for four samples, using the assumptions discussed previously. The weight per cent Fe in the olivine and pyroxene were taken to be 20% and 10% respectively. Other minerals correspond to their ideal chemical formula except for the iron spinel which is assumed to have 60% Fe. The very small peak areas are considered to be upper limits because of their very large statistical errors, and the other values will have semiquantitative significance. However, the relative values for one mineral from sample to sample should be much more accurate.

For example, the iron percentage increases in the order 10 044 < 10 045 (crystalline rocks) < 10 060 (microbreccia) < 10 084 (fines). Thus the iron content increases on breakdown of the crystalline rocks. It might be reasonable to suppose that the iron was formed by reduction of FeS, but the increase in iron content is not parallelled by a decrease in troilite content. This would suggest that the iron is of meteoritic origin.

These spectra indicate the very large ilmenite content of these rocks—much larger than terrestrial basalts, and strongly suggest that the moon did not break away from the earth. It is also obvious that the ilmenite content in the fines and microbreccia is substantially lower than that in the crystalline rocks.

The spectra of the samples from the Apollo 12 mission are qualitatively similar. The ilmenite (peaks 1 and 1') and iron (peaks 6, 6' and 6") content is much lower in the Apollo 12 fines than in the Apollo 11 fines.

Table 9.3 Percentages of iron-bearing minerals in lunar samples

Mineral	% of total Mössbauer area					Weight % of total sample			
	10084	10045	10044	10060	10084		10045	10044	10060
Ilmenite	19.7	26.9	25.1	20.1	4.2		7.4	5.8	4.3
Pyroxene + glass	67.6	60.8	70.4	74.3	53.1		60.6	59.2	58.3
Olivine	4.4	6.1			1.8		3.0		
Iron	5.8	≤2.1	≤1.6	3.2	0.6		≤0.2	≤0.1	0.3
Troilite	≤1.1	≤2.0	≤1.5	≤1.0	≤0.1		≤0.2	≤0.2	≤0.1
Iron spinel	≤1.4	≤2.1	≤1.4	≤1.4	≤0.2		≤0.2	≤0.2	≤0.1
Non-iron					40.0		28.4	34.5	36.9

P. Gay, G. M. Bancroft and M. G. Bown, *Proc. Apollo 11 Lunar Sci. Conf. Geochim. Cosmochim. Acta Suppl. 1*, **1**, 481, Pergamon.

9.4 Terrestrial samples and a potential use in geochemical prospecting

Previous sections indicate that about 0.1 wgt % of a magnetic mineral (Table 9.3) may be detected using Mössbauer techniques, while somewhat less of a paramagnetic mineral could be detected if it has peaks well separated from other paramagnetic minerals. With improvement in techniques it should be possible in the next decade to extend the detection range down to 0.01%. At the above range, it could be very difficult to detect minerals using other standard techniques, even if the mineral is well crystallized.

It would appear then that the Mössbauer effect may be very useful to the geochemical prospector for first, rapid semiquantitative modal analysis of rocks as described in previous sections of this chapter; and second, detection of small amounts of iron ore minerals formed a long way from the actual ore body but indicative of ore reactions. Such reaction aureoles could be monitored by taking Mössbauer spectra of selected samples, possibly enabling the ore body to be more rapidly pinpointed.

An example of this latter type of potential use for the Mössbauer effect can be illustrated by using known chemical reactions between silicates and sulfur. Sulfur reacts with olivine as follows:

$$2Fe_2SiO_4 + S \longrightarrow FeS + Fe_3O_4 + 2SiO_2 \qquad (9.3)$$

or

$$2Fe_2SiO_4 + 2S \longrightarrow FeS_2 + Fe_3O_4 + 2SiO_2 \qquad (9.4)$$

or

$$3Fe_2SiO_4 + 4S \longrightarrow 2FeS_2 + 2Fe_2O_3 + 3SiO_2. \qquad (9.5)$$

In a similar fashion, hedenbergite reacts with sulfur:

$$4CaFeSi_2O_6 + S \longrightarrow FeS + Fe_3O_4 + 4CaSiO_3 + 4SiO_2. \qquad (9.6)$$

The sulfur may have been in the form of gaseous H_2S or SO_2 which diffused a long way from the place of the main reaction, and reaction aureoles of FeS, FeS_2 or Fe_3O_4 may be present a good distance from the main ore body. The Mössbauer effect would seem to be very useful here for recognizing such reaction aureoles using the known spectra of FeS and Fe_3O_4.

From a longer range viewpoint, the development of rapid and precise thermometry measurements using the site populations in silicates (chapter 8) could be of the greatest potential use to the mining engineer. Temperature gradients deduced from thermometry measurements, when combined with controlled laboratory studies can be indicative of the direction and distance of important mineralization. In addition, once an ore body is located, temperature gradients can be indicative of decreasing temperatures of the ore solutions with increasing distance from their sources, and hence of the direction of movement of such solutions.

References

9.1 C. E. Johnson and G. P. Glasby, *Nature*, **222**, 376 (1969).
9.2 E. L. Sprenkel-Segel and S. S. Hanna, *Geochim. Cosmochim. Acta.*, **28**, 1913 (1964); ibid., *Mössbauer Effect Methodology*, Ed. I. J. Gruverman, Plenum Press **2**, 89 (1966).
9.3 C. L. Herzenberg, *Mössbauer Effect Methodology*, ed. I. J. Gruverman, Plenum Press **5**, 209 (1970).

Bibliography

A. M. Reid and K. Fredericksson, Chondrules and Chondrites, in *Researches in Geochemistry*, Abelson ed., Wiley and Sons (1967), vol. 2.
C. L. Herzenberg, *Mössbauer Effect Methodology*, Ed. I. J. Gruverman, Plenum Press, **5**, 209 (1970).

Appendix 1. Energy conversion tables and important physical constants

	cm^{-1}	joule	eV	$Mc\,s^{-1}$†
$1\,cm^{-1}$	1	1.9862×10^{-23}	1.2398×10^{-4}	2.9979×10^{4}
1 joule	5.0348×10^{22}	1	6.2420×10^{18}	1.5094×10^{27}
1 eV	8066.0	1.6020×10^{-19}	1	2.4181×10^{8}
$1\,Mc\,s^{-1}$	3.3356×10^{-5}	6.6252×10^{-28}	4.1355×10^{-9}	1

† For conversion of $Mc\,s^{-1}$ to $mm\,s^{-1}$, see Appendix 3.

Planck's constant $h = 6.626 \times 10^{-27}$ erg-sec $= 6.626 \times 10^{-34}$ joule-sec
Boltzmann's constant $k = 1.380 \times 10^{-16}$ erg/K $= 1.380 \times 10^{-23}$ joule/K
Avagadro's number $N = 6.025 \times 10^{23}$ atoms/mole
Electron charge $e = 4.803 \times 10^{-10}$ e.s.u. $= 1.602 \times 10^{-19}$ coulombs
Electron mass $m_e = 9.108 \times 10^{-28}$ g $= 9.11 \times 10^{-31}$ Kg.
Velocity of light $c = 2.998 \times 10^{10}$ cm s^{-1} $= 2.998 \times 10^{8}$ m s^{-1}

Appendix 2. Velocity conversions for Fe and Sn Mössbauer data

(1) *Fe data*

Source	To convert to $Na_2[Fe(CN)_5NO].2H_2O$, add: (mm s^{-1})
Pt	+0.607
Cu	+0.483
Pd	+0.442
Fe	+0.257
$K_4Fe(CN)_6.3H_2O$	+0.215
Stainless steel	+0.16
Cr	+0.075
$Na_2[Fe(CN)_5NO].2H_2O$	0.000

(2) *Sn data*

Source	To convert to $SnO_2(BaSnO_3)$, add: (mm s^{-1})†
Mg_2Sn	+1.82
β Sn	+2.70
SnO_2	0
α Sn	+2.10
Pd/Sn	+1.52
$BaSnO_3$	~0†

† It is normally assumed that SnO_2 and $BaSnO_3$ have identical centre shift values. No accurate measurement of any difference in centre shift values has been yet performed.

Appendix 3. Conversion of mm s^{-1} to Mc s^{-1}

From eq. 1.14, $\Delta E = \dfrac{v}{c} E_\gamma$

For ^{57}Fe, $E_\gamma = 14\cdot39$ keV

$$\text{For } v = 1 \text{ mm s}^{-1}, E = \frac{14\cdot39 \times 10^3}{2\cdot998 \times 10^{11}} = 4\cdot800 \times 10^{-8} \text{ e.v.}$$

$4\cdot800 \times 10^{-8}$ e.v. $\times 1\cdot602 \times 10^{-19}$ coulombs $= 7\cdot690 \times 10^{-27}$ joules
$\hspace{8cm} = 7\cdot690 \times 10^{-20}$ ergs

From $E = h\nu$

$$\nu = \frac{7\cdot690 \times 10^{-20}}{6\cdot625 \times 10^{-27}} = 11\cdot61 \text{ Mc s}^{-1}$$

Thus for ^{57}Fe, 1 mm s^{-1} = 11·61 Mc s^{-1}.

For other isotopes having a γ ray energy E_γ(keV)

$$1 \text{ mm s}^{-1} = \left(11\cdot61 \times \frac{E_\gamma}{14\cdot39}\right) \text{Mc s}^{-1}$$

For the isotopes discussed in this book:

	E_γ (keV)	1 mm s^{-1} = (Mc s^{-1})
^{57}Fe	14·39	11·61
^{99}Ru	90	72·59
^{119}Sn	23·88	19·26
^{121}Sb	37·15	29·96
^{125}Te	35·48	28·62
^{129}I	27·75	22·38
^{129}Xe	39·58	31·92
^{193}Ir	73·08	58·96

Answers

Chapter 1
1. 4.9×10^{-11} eV $(4.0 \times 10^{-7}\,\text{cm}^{-1})$; 2.4×10^{-8} eV $(1.9 \times 10^{-4}\,\text{cm}^{-1})$; 4.9×10^{-6} eV $(0.04\,\text{cm}^{-1})$.
2. $E_R = 6.97 \times 10^{-2}$ eV (Zn), 2.58×10^{-3} eV (Sn), and 4.72×10^{-2} eV (Ir); $E_D = 8.48 \times 10^{-2}$ eV (Zn), 1.63×10^{-2} eV (Sn), and 6.98×10^{-2} eV (Ir).
3. $v = 4.48 \times 10^5$ mm s^{-1} (Zn), 6.48×10^4 mm s^{-1} (Sn), 2.19×10^5 mm s^{-1} (Ir). These velocities are very difficult to obtain.
4. $E_R = 2.81 \times 10^{-14}$ eV. This is orders of magnitude less than the linewidths.
5. $v = 1.6 \times 10^{-4}$ mm s^{-1} (Zn), 0.31 mm s^{-1} (Sn), 1.0×10^1 mm s^{-1} (Ir). Note—the linewidths observed are twice these i.e. source + absorber linewidth. The scan would be typically ± 5 to ± 10 times the linewidth. The velocity for Zn and Ir are not easy to obtain (too low and too high respectively).
6. $f = 0.96$.
7. $f = 0.03$. Compared to the 14.4 keV gamma ray, the 136.3 keV gamma will give exceedingly poor spectra.

Chapter 2
1. Tetrahedral, $q_{\text{lattice}} = 0$; square planar, $e^2 q_{\text{lattice}} Q(1 - \gamma) = +2.6$ mm s^{-1}. q_{valence} would give large temperature dependent Q.S. for tetrahedral; smaller and temperature independent negative Q.S. for square planar.
2. See reference.
3. f factor will be extremely small, $E = 0.134$ eV, $\Gamma_{\text{source}} = 0.14$ mm s^{-1}, minimum line width in Mössbauer experiment = 0.28 mm s^{-1}. If $e^2 qQ_{\text{ex}} = 3$ mm s^{-1}, separation of peaks = 0.21 mm s^{-1}, 0.42 mm s^{-1} and 0.84 mm s^{-1} and resolution will be relatively poor.
4. See reference.
5. M. G. Clark, G. M. Bancroft and A. J. Stone, *J. Chem. Phys.*, **47**, 4250 (1967).
6. cis isomer, Q.S. = 0; trans isomer, $\eta = 1$, sign indeterminate.
7. B. A. Goodman and N. N. Greenwood, *Chem. Comm.*, 1105 (1969); N. E. Erickson, *Chem. Comm.*, 1349 (1970).

Mössbauer spectroscopy

8. T. C. Gibb and N. N. Greenwood, *J. Chem. Soc.*, 6985 (1965).
9. M. Pasternak and T. Sonnino, *J. Chem. Phys.*, **48**, 2009 (1968). $\delta_1 = 1.5$ cm s^{-1}, $\delta_2 = 3.0$ cm s^{-1}, $\delta_3 = 0.6$ cm s^{-1}, $\delta_4 = 1.2$ cm s^{-1}, $\delta_5 = 1.8$ cm s^{-1}, $^gE_4 = +1.5$ cm s^{-1}, $^gE_3 = +0.9$ cm s^{-1}, $^gE_2 = -0.3$ cm s^{-1}, $^gE_1 = +1.5$ cm s^{-1}; $^eE_3 = +2.0$ cm s^{-1}, $^eE_2 = +0.5$ cm s^{-1}, $^eE_1 = -2.5$ cm s^{-1}, C.S. $= +0.1$ cm s^{-1}. For Q.S., note that conversion to Mc s^{-1} and ^{127}I is necessary—see reference.

Chapter 3

1. No, because the velocity per channel would be about 0.1 cm s^{-1}, and the great majority of the γ rays going into the resonant channel would be non-resonant photons. Greater than 4000 channels would be desirable.
2. Velocity per channel = 0.03114 mm s^{-1}.
   Zero velocity with respect to source = 127.24 channels.
   Centre shift of Fe = -0.194 mm s^{-1}. (Quoted as -0.185 in literature.)
3. This small difference in χ^2 cannot be taken as conclusive. Spectra at other temperatures are desirable and X-ray results would be useful.

Chapter 4

1. See J. Knight and M. J. Mays, *J. Chem. Soc.* (A), 654 (1970).
2. Just a single doublet in solution. Seven coordinate Fe. Quadrupole splitting similar to Fe$_A$ in Fig. 4.3a.
3. N. E. Erickson and N. Sutin, *Inorg. Chem.*, **5**, 1834 (1966).
4. M. B. Robin and P. Day, *Adv. Inorg. Radiochem.*, **10**, 247 (1967); A. K. Bonnette and J. F. Allan, *Inorg. Chem.*, **10**, 1613 (1971).
5. Most likely FeCl$_2$dppe(tetrahedral); magnetic susceptibilities.

Chapter 5

1. (a) FeBr$_3$ < FeBr$_2$ ($\delta R/R$ is $-$ve, remove d electron from Fe^{2+} to Fe^{3+}).
   (b) SnBr$_4$ < SnBr$_2$ ($\delta R/R$ is $+$ve, remove s electrons from Sn^{2+} to Sn^{4+}).
   (c) AuBr < AuBr$_3$ ($\delta R/R$ is $+$ve).
   (d) SbBr$_3$ < SbBr$_5$ ($\delta R/R$ is $-$ve; compare with SnBr$_2$, SnBr$_4$).
   (e) Fe(CO)$_5$ < Fe(CO)$_4$PPh$_3$.
   (f) $L = $ N$_2$ < $L = $ (CH$_3$)$_2$CO ($\sigma + \pi$ for N$_2$ should be larger than for (CH$_3$)$_2$CO).
   (g) Cl$_3$SnMn(CO)$_5$ > Ph$_3$SnMn(CO)$_5$.
2. (a) 0.50 mm s^{-1} (b) 0.30 mm s^{-1} (c) ≤ 0.50 mm s^{-1}
   (d) 0.39 mm s^{-1} (e) 0.24 mm s^{-1}
3. R. V. Parish, *Prog. Inorg. Chem.*, **15**, 101 (1972), page 145.

Chapter 6

1. Ph$_4$Sn and Cl$_4$Sn, Q.S. = 0; Q.S. (Ph$_3$SnCl) = Q.S. (PhSnCl$_3$) = 1/1.15 $\times$ Q.S. (Ph$_2$SnCl$_2$).

Answers

2. (a) tetrahedral has Q.S. = 0; square planar would have large negative Q.S. ($Q = -\text{ve}$).
   (c) because of large $q_{valence}$, *trans* will have *smaller* Q.S. than *cis*.
   (e) the two will differ by ratio of Q and q_{5p} values.
3. Ideal $SnCl_3^-$, Q.S. = 0. The quadrupole splitting increases, then decreases as the tetrahedral angle is approached; the centre shift decreases continuously.
4. Five geometric isomers. G. M. Bancroft and E. T. Libbey, *J. Chem. Soc.* (Dalton). In Press (1973).
5. For the 3 MAB_3C_2 isomers, the calculated quadrupole splittings are -0.80 mm s^{-1}, $+1.18$ mm s^{-1} and $+0.44$ mm s^{-1}. For the five $MA_2B_2C_2$ isomers, the calculated quadrupole splittings are -1.10 mm s^{-1}, $+1.40$ mm s^{-1}, -0.70 mm s^{-1}, $+1.30$ mm s^{-1} and -0.20 mm s^{-1}. When (p.q.s.)$_A$ = (p.q.s.)$_B$, all isomers give either the *cis*- or *trans*-MC_2B_4 value of -0.50 mm s^{-1} or $+1.00$ mm s^{-1}.
6. G. M. Bancroft and R. H. Platt, *Adv. Inorg. Radiochem.*, **15**, 59 (1972) Academic Press Table 27.
7. -0.062 barns.
8. $+0.28$ mm s^{-1}, yes.
9. Q.S. and C.S. for dppe compounds should be 0.24 mm s^{-1} more negative and 0.08 mm s^{-1} more positive than their depe analogues. C.S. should vary in the order $(CH_3)_2CO > NH_3 \sim N_2 > CO$. Q.S.:CO (most negative) $< NH_3 < (CH_3)_2CO \sim N_2$.
10. Tetrahedral: $+2.54$ mm s^{-1}.
    Octahedral: $+1.90$ mm s^{-1}, indicating octahedral structure.
11. Becomes less negative. B. W. Dale and R. V. Parish, *J. Chem. Soc.* (Dalton). In Press (1973).

Chapter 7

1. See reference.
2. S. S. Hafner and H. G. Huckenholtz, *Nature*, **233**, 9 (1971).
3. Inner two and outer two. Otherwise the centre shifts are not reasonable. Spectra at other temperatures would give better resolution. Peaks are too close together to resolve properly.
4. D. L. Smith and J. J. Zuckerman, *J. Inorg. Nucl. Chem.*, **29**, 1203 (1967).

Chapter 8

1. See Table 8.4.
2. It is debatable whether this evidence for ordering is any more significant than that given by Burns. H_{B_1}/H_{B_2} are usually ~ 0.90. Widths are equal within errors.
3. Mössbauer spectra distinguish between Fe and Mn.

Index

Absorbers, critical, 58
  preparation of, 52–53, 83
Acmite-jadeite, Mössbauer parameters of, 169, 190
  structure of, 167–168
Actinolites, 177
  Mössbauer parameters of, 182, 191–193
  spectra of, 182–184
  structure of, 179
Activation energies for order–disorder, 218–219
Alkali Amphiboles, colour and pleochroism, 194–196
  Mössbauer parameters of, 182, 191–193
  optical spectra of, 194–196
  oxidation of, 196
  site populations in, 201, 206
  spectra of, 182–184
Allanite (see epidotes)
Almandine garnet, 158, 162
α emission, 2–3
  in americium-241, 3–4
Americium-241, 3
Amosite (see cummingtonite-grunerite)
Amphiboles (see also individual minerals)
  in bulk samples, 227
  cation-oxygen bond lengths in, 179
  chemical formula of, 177
  charge transfer spectra in, 190–193
  colour and pleochroism in, 194–196
  crystal field stabilization in, 223
  end members of, 177
  line widths in, 189
  oxidation and weathering in, 196–197
  parameters of, 182–184
  site populations in, 201–212
  structural variations and quadrupole splittings in, 190–193
  structure of, 178–180
  structure and spectra of, 62–64, 180–184
Amplifiers, 47, 50–52
Anthophyllites, Mössbauer parameters of, 182
  spectra of, 62–64, 180

Antimony-121, 42, 121
  Sb^{III} compounds, 99–100
  calibration for, 55
  centre shifts in, 99–100
  centre shift and oxidation state in, 71
Area ratio method, 202–238
  accuracy of, 204–206
  for amphiboles, 206–212
  assumptions in, 202–205, 228–229
  and computing, 204
  ferric to ferrous ratios using, 205–207
  in lunar samples, 220, 234–238
  in meteorites, 229–234
  for multiphase assemblages, 226–238
  for pyroxenes, 212–214
  and saturation, 203
Areas (see also area ratios), 4, 17, 41, 69, 72, 158–159, 171, 173, 176, 183, 189–190
  and f factors, 41
  in lunar samples, 235–236
  and saturation, 41
Asymmetry parameter, η, 23–27, 34–35, 40, 112–114, 134–136
Atomic number, 2
Auger electrons, 3, 86–88
Augites, line broadening in, 172–176, 189–190
  in lunar samples, 234–237
  Mössbauer parameters of, 169
  site populations in lunar samples, 220
  spectra of, 171–177
  structure of, 166, 168

Background, 57–58
  and f factors, 58
  ^{119}Sn spectra, 58
β emission, 2–3, 86–88 (see also ^{57}Co)
  after-effects, see γ emission
  electron capture, 3, 86–88
  negatron decay, 2
  positron decay, 2, 3

Calibration, 53–57
  and Fe metal, 55

Index

Calibration—*continued*
  and laser interferometer, 55–57
  and scan centre, 55
  and velocity conversions, 241
Centre shift (*see also* isomer shift, partial centre shift, and individual elements)
  in antimony-121, 97–100
  and bonding properties of ligand, 94–96, 102–108, 116–120, 143–145
  and coordination number in minerals, 162–165
  and correlation with quadrupole splittings, 113–115, 118–121, 143–144, 150–151
  electronegativity variations of, 91–94, 98–101
  fingerprint uses of, 68–83, 156–191
  in gold-197, 97, 107
  in iodine-129, 73–74, 91–93
  in iron-57, 21, 68–73, 75–83, 94–96, 103–108, 156–190
  and oxidation states, 68–71, 156–162
  in ruthenium-99, 96–97, 107
  in tin-119, 21, 71–72, 83–84, 97–99, 101–103, 162–165
  and s-electron density at the nucleus, 90–102, 113–116
  and second order Doppler shift, 17–18
Cesium-133, 42
Charge-transfer spectra, 194–196
χ^2, 59–66, 160, 164
Cobalt-59, 121, 131–132
Cobalt-57, 15, 47
  decay of, 3
  emission spectra using, 86–88
  half-life of, 4, 42–43
  production of, 3
  sources, 19, 50, 52
Cobalt-61, 43
Coincidence techniques, 58
Collimation of γ beam, 53
Computational methods, 58–66
  and χ^2 values, 59–66
  and a 'good' fit, 60–62
  and Lorentzian lines, 59–66
  and mineral spectra, 62–66
Constraints (in computing), 60, 62–66, 155, 189–190
Coordination numbers and Mössbauer spectra, 156, 162–164, 167
Copper-61, 2, 43
Cosine effect, 53–54
Critical absorber, 58
Crocidolite (*see* alkali amphiboles)
Crystal-field stabilization energies, 221–224
Cummingtonite–Grunerite, crystal field stabilization in, 223
  linewidths in spectra of, 189

Cummingtonite–Grunerite—*continued*
  Mössbauer parameters of, 182, 191–193
  oxidation of, 196–197
  site populations in, 202–212
  spectra of, 180–181
  structure of, 178–179

Deerite, 161–162, 189
Diopside–Hedenbergite, Mössbauer parameters of, 169, 172, 191–193
  structure of, 165–168
Doppler shifts, 1, 7–9, 13, 17, 48, 53–54
Doppler widths, 11–13
Drive mechanisms, 47–49

Einstein model of solid, 14
Electric field gradient (q), 7, 22, 23–27 (*see also* quadrupole splitting)
  and q_{lattice}, 27–38, 31–32, 110, 147–149, 190–191
  and q_{valence}, 27–29, 32, 110, 147–149, 190–191
  Jahn–Teller distortions, effect on, 31
  orbital populations in, 28, 29, 112–113
  separation into $q_{\text{M.O.}}$ and $q_{\text{C.F.}}$, 29–32, 110–121, 124–126, 145–152
  spin orbit coupling, 31, 146–149
  temperature dependence of, 31, 146–149
  Sternheimer factors in, 27
  units, 27
Electric field gradient tensor components, 23–26, 134–136
Electron binding energy, 4
Electron capture, 3 (*see also* β emission)
  and Auger electrons, 3, 86–88
Electronegativity, and anomalous Sn^{IV} centre shifts, 101–102
  centre shift correlations with, 99–111
  and centre shift, 90–94
  derivation of Pauling values, 98–99
  and quadrupole splittings in ^{129}I and ^{129}Xe compounds, 112–116
Electron probe, 176, 204
Emission spectra, 86–88
Epidotes, 158, 162, 187–190
Europium-151, 71
Exchange reaction, cations in minerals, 215–221

Fassaite, 165
Frozen solution studies, 83–85

γ emission, 2–5, 8–15, 17 (*see also* cobalt-57)
  and background, 5–7
  detectors and electronics for, 50–52
  emission spectra, 87

Index

γ emission—*continued*
   and Heisenberg uncertainty principle, 4
   and internal conversion, 4
   linewidths of, 4–5
   nuclear gamma resonance, 8–15
   orientation of sample using, 33–34
   sources for, 50
   statistics in, 58
   and a useful Mössbauer isotope, 41–42, 47
Garnets, 158, 163
Gedrite–Ferrogedrite, 177
Geochemical prospecting, 238
Geothermometer, 193, 201, 215–220
   and geochemical prospecting, 238
   and kinetics of cooling, 217–219
   and lunar samples, 220
   using orthopyroxenes, 212–220
Germanium-73, 42
Gillespite, 158, 161–162, 191
Glasses, 83–85
Glaucophane (*see* alkali amphiboles)
Gold-197, 42, 43
   centre shift and oxidation state in, 71
   centre shift and partial centre shift in, 104
   and the spectrochemical series, 107
   and Au^{III} compounds, 104, 151–152
   quadrupole splittings in, 111, 145, 151–152
Goldanskii–Karyagin asymmetry, 34, 190
Greigite, 160

Hafnium-177, 42
Half-life of excited states, 4–5, 87–88
Hedenbergite, 238 (*see also* diopside-hedenbergite)
Heisenberg line widths, 4–5, 11–15
   compared with E_γ and modulation energy, 7, 8
   and mean life of excited state, 4, 5
   table of, 42
Heisenberg uncertainty principle, 4–5, 7, 40
High pressure studies, 82–83, 156, 161–162
High spin–low spin in Fe, 80–82, 161–162
Hornblende, 177, 179, 182–184
Howieite, 159
Hypersthene, 165, 168 (*see also* orthopyroxenes)

Ilmenite, in lunar samples, 234–237
   weathering in, 198
Infrared spectroscopy, 1, 8–13
   site populations in silcates, 194, 201
Internal conversion, 4, 43
Iodine-129, 30, 42, 71
   alkali iodides, 91–93, 113
   calibration for, 55
   centre shifts in, 91–93, 102

Iodine-129—*continued*
   emission spectra using, 86–87
   fingerprint uses in, 73–75
   and iodine halides, 73–75, 91–93, 111–114
   quadrupole splitting in, 111–114
   spectra and derivation of parameters in, 135–138
Iridium-191, 13–14
Iridium-193, 42, 121
   centre shifts in, 71
   emission spectra of, 87
   quadrupole splittings in, 130–131
Iron-57, 1, 3, 15, 41–44 (*see also* iron compounds and minerals)
   areas, 40–41, 202–205
   calibration for, 53–57
   centre shifts in, 68, 70–71, 90–91, 94–96, 102–108
   computational methods for, 58–67
   cross section at resonance in, 40–42
   decay scheme for, 3
   detectors for, 51–52
   emission spectra from ^{57}Co, 86–88
   excited state of, half life, 4, 42
   γ ray, 14.4 keV, Doppler shift for, 8
   Heisenberg width, 5, 42
   Isomer shift in, 19–21
   Isotope fractionation, 203, 228
   Magnetic splitting in, 38–40
   Quadrupole splitting in, 21–23, 110–111
      calculation of, 32–33
      and distortion of mineral sites, 190–193
      and $q_{M.O.}$, 29–30, 116–120
      and $q_{C.F.}$, 30–31, 145–151
      sign of, 33–34
   saturation corrections for, 41, 202–203
   sources for, 49–50
Iron-56, 3
Iron (IV), 68, 78
Iron metal, 227–228, 234–237
Iron (I) high spin, 68, 158
Iron π-cyclopentadiene compounds, 34, 75, 76, 85, 150, 151
Iron sulphides, 160–162, 227–230, 234–238
Iron (VI), 68, 158
Iron (III) high spin, acetylacetonates, 82–83, 88
   amphiboles, *see* alkali amphiboles
   in bulk rock samples, 227–237
   centre shifts and oxidation state, 68, 156–162
   and computation in silicate minerals, 61–66
   in deerite, 161–162
   in epidotes, 158, 187–189
   ferric to ferrous ratios, 159–162, 205–207
      in colour and pleochroism, 194–196

Index

Iron (III) high spin—*continued*
    in oxidation and weathering, 196–198
    in halides, 158
    in howieite, 159–160
    iron oxides, 221–224, 227, 232–234
    in micas, 164, 184–187
    in orthoclase, 158
    oxalates, 77–79, 83
    in oxidized meteorites, 233–234
    pyroxenes, *see* pyroxenes
    quadrupole splittings, 121, 190
    in sapphirines, 163–164
    solid state reduction of, 77–79
    sulphides, 160
Iron (III) low spin, 29–31, 145, 148, 158
Iron (II) high spin, in amphiboles, (*see* alkali amphiboles and cummingtonite, anthophyllite)
    in bulk rock samples, 227–237
    centre shifts, 91, 100–101, 158
    centre shifts and oxidation state, 68, 156–162
    and computation in silicate minerals, 61–66
    crystal field splittings, 29–32, 147–149
    in deerite, 161–162
    energy level diagram, 146
    in epidotes, 187–189
    ferric to ferrous ratios, 159–162, 205–207
        and colour and pleochroism, 194–196
        and oxidation and weathering, 196–198
    ferrous or ferric, 156–158
    ferrous site populations, 201–220
    in garnets, 150, 162
    in gillespite, 158, 161–162, 191
    halides, 100–101, 103
    high spin–low spin, 80–81, 161–162
    in howieite, 159–160
    isoquinoline complexes, 148–149
    in lunar samples, 234–237
    in meteorites, 229–233
    in micas, 164, 184–187
    and nephelauxetic series, 108
    and partial centre shifts, 107
    olivines, 162, 223, 228–231, 234–238
    oxalates, 77–79
    phenanthroline compounds, 80–81
    pyroxenes, *see* pyroxenes
    quadrupole splittings and distortion, 190–192
    and $q_{C.F.}$, 29–32, 111, 145–150
    in sapphirines, 163–164
    spinels, 158, 162
    in staurolite, 158–162
    sulphides, 160, 227–230, 234–238
    tetrahedral $FeX_4^=$ species, 146–149
Iron (II) intermediate spin, 80–81, 88

Iron (II) low spin, bonding in, 94
    carbonyls, 71, 106, 143
    centre shifts, 95–96, 105–107, 158
    centre shift-quadrupole splitting correlations, 118–119
    and bonding properties of ligands, 118–119
    cyanides, $Fe(CN)_5X$, 95, 105, 116–117, 128
    diphosphine compounds, 96, 105, 116–118, 128
    high spin–low spin, 80–82
    isocyanides, 71, 104–105, 123, 128, 143
    niox compounds, $Fe(niox)_2YZ$, 119–120
    partial centre shifts, 104–107
        and correlation with spectrochemical series, 107–108
    partial quadrupole splittings, 127–130
        bonding, 143–145
        and Co^{III}, Ir^{III}, Ru^{II} quadrupole splittings, 130–133
        structure, 142–143
    quadrupole splittings, 116–119, 123, 128
    sulphides, 160
Iron (0) 29, 32, 69, 146
    partial centre shifts, 102
    polynuclear carbonyls, 72–77, 85
    quadrupole splittings, 150–151
Isobars, definition, 2
Isomer shift (*see also* centre shift and partial centre shift), 17–21
    and nuclear radius, 7, 18–19, 90–102
    and *s*-electron density at the nucleus, 90–102, 113–116
Isotopes, definition, 2

Johannesite, 165

Kamacite, 237
Kinetics of order and disorder, 217–219
Krypton-83, 42

Laser interferometer, 55–57
Line shape, Lorentzian, 4, 40–44 (and *see* below)
    and non-equivalence of peaks, 74–77
    and non-Lorentzian line shapes, 62–66
    and purity, 69
    and ratio method, 203–206
    and silicate minerals, 161, 189–190
    and sources, 50
    and thickness effects, 41
Line widths (or widths at half height), 4, 5, 9
    and absorber preparation, 52
    and computing, 59, 61–66, 204
    and Doppler widths, 11–13
    in lunar samples, 235–236

Index

Line widths—*continued*
  and non-equivalence in inorganic compounds, 74–76
  in orthopyroxenes, 215
  and purity, 69
  in silicate minerals, 189–190
  and sources, 50
  and thickness broadening, 40–41
Li drifted germanium counters, 43, 50–52
Lorentzian (*see* line shape)
Lunar samples, 234–237

Mackinawite, 160–161
Magnesioriebeckite–Riebeckite (*see* alkali amphiboles)
Magnetic field method for sign of quadrupole splitting, 33–34
Magnetic moment, 6, 38–39
Magnetic splitting, 5–6, 17, 38–40
  in bulk samples, 226–238
Manganese nodules, 226–227
Mass number, 2
Mean square vibrational amplitude, 14–15
Meteorites, 229–234
Micas, chemical formula of, 184
  Mössbauer spectra of, 164, 186
  structure of, 185
  weathering of, 197–198
Mirror image spectra, 48–49, 55
Mössbauer fraction f, 14–15, 34
  and area ratio method, 202–204, 209, 231
  expression for, 14
  and frozen solution studies, 83–85
  and a good Mössbauer isotope, 41
  and line width, 40
  saturation, 41
  and sources, 50
Multichannel Analyzer, 47–52, 59

Negatron decay, 2 (*see also* β emission)
Neptunite, 158
Neptunium-237, 3, 71
Nickel-61, 2, 42

Olivines, 162, 238
  crystal field stabilization in, 223
  in lunar samples, 234–237
  in meteorites, 229–231
  site populations in, 224
Omphacites, Mössbauer parameters of, 169
  spectra of, 65, 173, 176–177
  structure of, 166, 168
Ordering of cations, in amphiboles, 179–184, 195–196, 206–212
  and crystal field phenomena, 221–224
  and charge transfer spectra, 194–195
  by infrared spectroscopy, 183, 201

Ordering of cations—*continued*
  in micas, 184–186
  by Mössbauer spectroscopy, 170–171, 180–183, 184–188, 202–220
  and oxidation and weathering, 196 198
  in pyroxenes, 166, 171, 173, 206, 212–215
  in sapphirines, 164
  in spinels, 221–223
  by x-ray diffraction, 166, 171, 173, 178–180, 185, 201, 206
Orthoclase, 158, 163
Orthopyroxenes, and amphibole spectra, 180
  crystal field stabilization in, 233
  geothermometer using, 215–221
  in meteorites, 229–231
  Mössbauer parameters of, 169, 191–193
  site populations in, 204–206, 212–221
  spectra of, 170–171
  structure of, 167–168
Osmium-186, 42
Osmium-183, 87
Oxidation states from Mössbauer spectra, in amphiboles, 180–184
  and charge transfer spectra, 194–196
  in epidotes, 187–189
  in iron compounds, 70–71, 77–81
  in iron minerals, 156–162
  in lunar samples, 234
  in micas, 186
  in pyroxenes, 166–177
  in tin compounds, 71
  and weathering, 196–198, 233–234
Oxidation and weathering, 196–198
Oxyamosite, 196

Partial centre shifts, 102–107
  assumptions in, 103
  bonding properties from, 106
  and correlation with spectrochemical series, 108
  and Au^{III} compounds, 104
  and Fe^{II} high spin compounds, 103
  and Fe^{II} low spin compounds, 104, 106
  and Sn^{IV} compounds, 102
Partial field gradients, 25–26, 122–139
Partial quadrupole splittings, 111, 121–145
  assumptions in, 122
  bonding of ligands from, 126
  bonding in Fe^{II} low spin compounds, 143–145
  derivation of, for Fe^{II} low spin compounds, 127–130
  for Sn^{IV} compounds, 137–139
  molecular orbital approach in, 124–125
  point charge approach for, 122
  prediction of, Fe^{II} quadrupole splittings using, 130, 143

Index

Partial quadrupole splittings—*continued*
  Sn^{IV} quadrupole splittings using, 133, 142
  signs of Co^{III} and Ru^{II} quadrupole splittings using, 130–132
  structural uses for Sn and Fe compounds, 139–143
  and 2: −1 *trans-cis* ratio, 122–123
  use in calculating $Q_{^{57}Fe}$, 131–132
Pentlandite, 160–162, 164–165
π acceptance of ligands, and centre shifts in Fe^{II}, Ru^{II} compounds, 94–97
  and centre shifts and oxidation state, 71
  of neutral ligands, 117–121
  of N_2, 118
  and partial centre shifts, 106
  and partial quadrupole splittings, 126, 143–145
  and π-Cp Fe compounds, 151
  and quadrupole splittings, 111
    of Fe^{II}, Ru^{II} compounds, 29–30, 116–121
  and spectrochemical series, 107–108
  and 2: −1 *trans-cis* quadrupole splitting ratio, 124–126
Pigeonite, 165, 168, 191
Platinum-195, 42–44
Pleochroism, 156, 194–196
Positron decay, 2 (see also β emission)
Preamplifiers, 47, 50–52
Proportional counters, 43, 50–52
Pyroxenes (*see also* individual minerals)
  and amosite decomposition, 197
  anomalous spectra of, 173–177
  in bulk samples, 227–233
  cation-oxygen bond lengths in, 168
  chemical formulae of, 165
  crystal field stabilization in, 223
  end members of, 166
  geothermometer using, 215–221
  kinetics of ordering, 217–221
  linewidths in spectra of, 189
  in lunar samples, 220, 234–237
  ordering of cations in, 201–206, 212–221
  spectra and parameters of, 64–66, 169–177
  structural variations and quadrupole splittings in, 191–193
  structure of, 165–168
  structure and spectra of, 65, 167–169
Pyrrhotite, 160

Quadrupole moments, 6, 23, 33, 35, 38, 42, 131–132
Quadrupole splittings, 7, 17, 22–38, 39–40, 44, 68, 110–154
  (*see also* electric field gradient, partial quadrupole splitting, and individual elements)
  angular dependence of spectra, 33–34

Quadrupole splittings—*continued*
  bonding and structure from, 110–152
  and Clebsch–Gordon coefficients, 35–36
  in cobalt-59, 121, 131–132
  correlation with centre shifts, 113–115, 118–121, 143–144, 150–151
  derivation of, for ^{129}I, 36–38
  diagnostic use of, 71–77, 80–87, 156–191
  in gold-197, 151–152
  Hamiltonian for, 35
  iodine-129, 22, 35–38, 111–114
  in iridium-193, 131
  in iron-57, 22, 28, 33–34, 39, 116–120, 127–130, 146–151, 156–191
  in manganese-55, 121
  in ruthenium-99, 120–121, 130
  signs of, 23, 28–31, 33, 110–121, 123, 129–143, 148–152
  in tellurium-125, 121
  in tin-119, 22, 33–34, 133–142
  in tungsten-182, 121
  units of, 32

Radius, nuclear and $\delta R/R$ (*see also* isomer shift and centre shift), 7, 18–21, 90–102
Recoil energies, 8–15
Rhenium-187, 42, 44
Ruthenium-99, 30, 42, 211
  centre shift and oxidation state, 71
  centre shift and bonding properties, 96–97
  and the spectrochemical series, 107
  quadrupole splitting and bonding properties, 116, 120–121, 130
  Ru^{II} compounds, 96–97, 120–121

Sapphirines, 163–164
Saturation, 41
  and site populations in silicates, 202–203, 211
s-character of Sn–L bonds, 101–102, 145
Scintillation counters, 43, 50–52
s-electron density at the nucleus ($[\Psi(0)_s]^2$)
  (*see* isomer shift and centre shift), 18–21
  and bonding properties of ligands, 90–102, 113–116
σ donation of ligands, and centre shifts in Fe^{II}, Ru^{II}, Au^{I} compounds, 94–97
  and Au^{III} compounds, 151–152
  of neutral ligands, 117–121
  of N_2, 118
  and partial centre shifts, 106–107
  and partial quadrupole splittings, 126, 143–145
  and quadrupole splittings, 110–111
    of Fe^{II}, Ru^{II} compounds, 116–121
  and 2:1 *trans-cis* quadrupole splitting ratio, 126, 143–145

Index

Single channel analyzer, 47, 50–52
Single crystal method for sign of quadrupole splitting, 33–34, 122–123, 145, 148–149
Silver-107, 42
Site populations (*see* ordering of cations)
Solid state decompositions, 77–79, 196–198
Spectrochemical series, and correlation with Fe^{II} centre shifts, 107–108
  and high spin–low spin in Fe^{II}, 80–81
Spinels, 158, 162, 197
  cation ordering in, 221–223
Spodumene, 165–166
Stacking processes, in solids, 84
Standard deviation of a count, 57–59
Staurolite, 158, 162

Tantalum-181, 42, 44
Tantalum-182, 2
Tellurium-125, 30, 42, 71, 121
Tin-118, 50
Tin-119, 1, 30, 42
  background corrections for, 58
  centre shifts in, 97–99, 101–102 (*see* isomer shift)
  fingerprint uses of, 71–72
  four-coordinate Sn^{IV} compounds, 83–84, 99, 101–102, 133, 141
  frozen solution spectra of, 83–85
  isomer shifts in, 19, 21
  partial centre shifts in, 102–103
  partial quadrupole splittings, and alkyl tin halides, 113, 136–142
  and four versus five coordination, 83–84, 111, 133–142
  structural predictions using, 133–142
  and tin acetates, 140

Tin-119—*continued*
  quadrupole splittings in, 22, 23, 24, 42–44, 127, 133–142
  sources for, 50
  and stannous fluoride, 72
  and tin hexahalides, 97–98, 102–103
  and tin trichloride as a ligand, 71
Tripuhyite, 156
Troilite, 160, 227–230, 238
  in lunar samples, 234–237
Tschermakite–Ferrotschermakite, 177
Tungsten-182, 2, 42, 51, 71, 121

Ultraviolet spectroscopy, 12–13

Vibrators, 8, 47–49
Visible spectroscopy, 194, 222–223
Vivianite, 196

Wustite, 162

Xenon-129, 30, 42
  centre shifts in, 93–94, 102
  centre shift and oxidation state in, 71
  emission spectra of, 86–87
  quadrupole splittings in, 114–116
  and xenon halides, 86–87, 93–94, 114–116
X-ray structures, of amphiboles, 177–179, 191–192, 195
  of epidotes, 187–188
  of iron carbonyls, 72–73, 75–77
  of micas, 184–185
  of pyroxenes, 165–167, 191–192
  and structural variations in silicates, 191
  of tin compounds, 139–142

Zinc-67, 42, 44